ELECTRICITY 4

ELECTRICITY 4

MOTORS, CONTROLS, ALTERNATORS

FOURTH EDITION

WALTER N. ALERICH

DELMAR PUBLISHERS INC.®

NOTICE TO THE READER

Delmar Staff:
  Administrative Editor: Mark W. Huth
  Developmental Editor: Marjorie A. Bruce
  Production Editor: Carol A. Micheli

For information address Delmar Publishers Inc.
2 Computer Drive West, Box 15–015
Albany, New York 12212–5015

Printed in the United States of America
Published simultaneously in Canada
by Nelson Canada,
A division of International Thomson Limited

10 9 8 7 6 5 4 3 2

Library of Congress Cataloging-in-Publication Data

Alerich, Walter N.
  Electricity 4.

  Includes index.
    1. Electric motors, Alternating current. 2. Electric generators–Alternating current. 3. Electric controllers. I. Title. II. Title: Electricity four.
  TK2712.A55   1986   621.31   86–2006

ISBN 0-8273-2534-7
ISBN 0-8273-2535-5 (instructor's guide)

CONTENTS

PREFACE

The fourth edition of ELECTRICITY 4 has been updated to reflect current devices, equipment, and techniques used in the installation of alternating current rotating machinery. At the same time, the text has retained the features that have made it so popular through previous editions.

Building upon the principles introduced in ELECTRICITY 3, the text covers a number of common controllers for ac machinery. Each type of control presented is thoroughly described and well-illustrated. A detailed explanation of the operation is provided and numerous typical schematic and wiring diagrams are given to familiarize students with common installations. This type of thorough explanation better prepares students to perform effectively on the job in the installation, troubleshooting, repair, and service of rotating machinery and its associated controls.

The knowledge obtained by a study of this text permits the student to progress to further study. It should be realized that both the development of the subject of electricity and the study of the subject are continuing processes. The electrical industry constantly introduces new and improved devices and materials, which in turn often lead to changes in installation techniques. Electrical codes undergo periodic revisions to upgrade safety and quality in electrical installations.

The text is easy to read and the topics are presented in a logical sequence. However, the instructor can elect to follow a different sequence depending upon available time and the requirements of individual programs.

Each unit begins with objectives to alert students to the learning that is expected as a result of studying the unit. An Achievement Review at the end of each unit tests student understanding to determine if the objectives have been met. Note that the problems presented in this text require the use of simple algebra only for their solution. Following selected groups of units, a summary review unit contains additional questions and problems that test student comprehension of a block of information. This combination of reviews is essential to the learning process required by this text.

All students of electricity will find this text useful, especially those in electrical apprenticeship programs, trade and technical schools, and various occupational programs.

It is recommended that the most recent edition of the National Electrical Code (published by the National Fire Protection Association) be available for reference as the student uses ELECTRICITY 4. Applicable state and local regulations should also be consulted when making actual installations. Features of the fourth edition include:

- Updated photos to reflect modern equipment, devices, and installations
- Modification of selected circuit diagrams to include solid-state devices
- Content updated to the requirements of the 1984 National Electrical Code
- New unit on electromechanical and solid-state relays
- New information on ac variable speed motor drives
- Addition of solid-state motor controllers
- New glossary

A combined Instructor's Guide for ELECTRICITY 1 through ELECTRICITY 4 is available. The guide includes the answers to the Achievement Reviews and Summary Reviews for each text and additional test questions covering the content of the four texts. Instructors may use these questions to devise additional tests to evaluate student learning. Student Study Guides to accompany each text will give students additional opportunities for classroom and laboratory practice.

ABOUT THE AUTHOR

Walter N. Alerich, BVE, MA, has an extensive background in electrical installation and education. As a journeyman wireman, he has many years of experience in the practical applications of electrical work. Mr. Alerich has also served as an instructor, supervisor and administrator of training programs, and is well-aware of the need for effective instruction in this field. A former head of the Electrical-Mechanical Department of Los Angeles Trade-Technical College, Mr. Alerich has written extensively on the subject of electricity and motor controls. He presently serves as an international specialist/consultant in the field of electrical trades, developing curricula and designing training facilities. Mr. Alerich is also the author of ELECTRICITY 3 and ELECTRIC MOTOR CONTROL, ELECTRIC MOTOR CONTROL LABORATORY MANUAL, and the coauthor of INDUSTRIAL MOTOR CONTROL.

ACKNOWLEDGMENTS

The revision of ELECTRICITY 4 was based on information and recommendations submitted by the following instructors:

Eugene Gentile and James Richards, Auburn Career Center, Painesville, OH 44077
Mark D. Kocher, Columbus Technical Institute, Columbus, OH 43162
James L. Gettings, Lyons Technical Institute, Philadelphia, PA 19134
Robert F. Smeal, Greater Johnstown Area Vocational Technical School, Johnstown, PA 15905
Richard M. Berube, Licking County Joint Vocational School, Newark, OH 43055
Michael F. Auth, Lyons Institute, Clark, NJ 07066
Sam A. Portaro, Davidson County Community College, Lexington, NC 27292
DeWitt Booth, Southeastern Community College, West Burlington, IA 52655
Charles W. Thompson, J.F. Drake State Technical College, Huntsville, AL 35811
Buck Deaver, Martin Community College, Williamston, NC 27892

The revised manuscript was thoroughly reviewed by the following instructors:

Gene A. Hilst, Blackhawk Technical Institute, Janesville, WI 53547
James Brozek, Brazosport College, Lake Jackson, TX 77566
Charles Lenau, Ranken Technical Institute, St. Louis, MO 63146
Ronald G. Oswald, Dean Institute of Technology, Pittsburgh, PA 15226

ELECTRICAL TRADES

The Delmar series of instructional material for the basic electrical trades consists of the texts, text-workbooks, laboratory manuals, and related information workbooks listed below. Each text features basic theory with practical applications and student involvement in hands-on activities.

ELECTRICITY 1
ELECTRICITY 2
ELECTRICITY 3
ELECTRICITY 4
ELECTRIC MOTOR CONTROL
ELECTRIC MOTOR CONTROL LABORATORY MANUAL
INDUSTRIAL MOTOR CONTROL
ALTERNATING CURRENT FUNDAMENTALS
DIRECT CURRENT FUNDAMENTALS
ELECTRICAL WIRING — RESIDENTIAL
ELECTRICAL WIRING — COMMERCIAL
ELECTRICAL WIRING — INDUSTRIAL
PRACTICAL PROBLEMS IN MATHEMATICS FOR ELECTRICIANS

EQUATIONS BASED ON OHM'S LAW

P = Power in Watts
I = Intensity of Current in Amperes
R = Resistance in Ohms
E = Electromotive Force in Volts

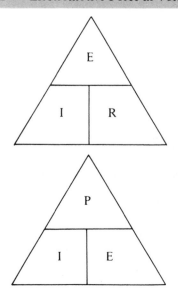

$$E = IR$$

$$I = \frac{E}{R}$$

$$R = \frac{E}{I}$$

$$P = IE$$

$$I = \frac{P}{E}$$

$$E = \frac{P}{I}$$

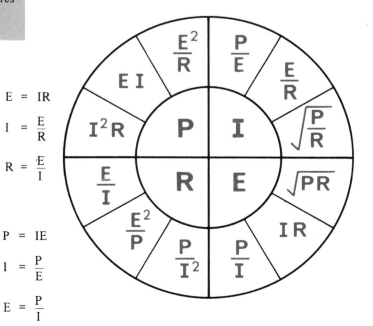

PHYSICAL AND ELECTRICAL CHARACTERISTICS OF THREE-PHASE ALTERNATORS

OBJECTIVES

After studying this unit, the student will be able to

- describe the purpose of an alternator.
- describe the ways in which the field of an alternator is established and how the alternator operates.
- explain the operation of the field discharge circuit.
- state how the frequency of an alternator can be determined and give the formula for calculating the frequency.
- explain how voltage control for an alternator is accomplished.
- describe the structure and operation of a rotating-field alternator.
- diagram alternator connections.
- explain three-phase voltages.

An *alternator* is a machine designed to generate alternating current (ac). This machine is the major electrical unit in power plants.

The alternator converts to electrical energy the mechanical energy of a prime mover such as a diesel engine, steam turbine, or water turbine. Another prime mover which has become increasingly important in the generation of electricity is the action of wind turning propellers.

THREE-PHASE VOLTAGES

Three phase is the most common polyphase electrical system. *Poly* means more than one. It is, in this instance, a system having three distinct voltages that are out of step with one another. There are 120 degrees between each voltage. Figure 1-1 shows sine waves taken on an electrical oscillograph instrument trace. This display shows the voltage relationships of the windings. This can be taken at any point in a three-phase system. The three phases are generated by placing each phase coil in the alternator 120 degrees apart, mechanically. A rotating dc magnetic field will then cut each phase coil in succession, inducing a voltage in each armature coil, out of step with each other. These armature coils may be connected internally or externally in a delta or a wye (or star) connection. Rotating fields are more commonly used than stationary fields because generating large amounts of current requires larger sizes of conductors and iron to rotate. Therefore, it is more practical to make the armature stationary.

Wye (star) and delta connections are shown in figure 1-2. These connections are shown in more detail in ELECTRICITY 3, under the heading of transformers.

1

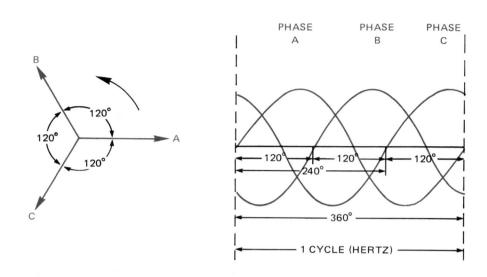

Fig. 1-1 Electrical displacement and generation of a three-phase voltage

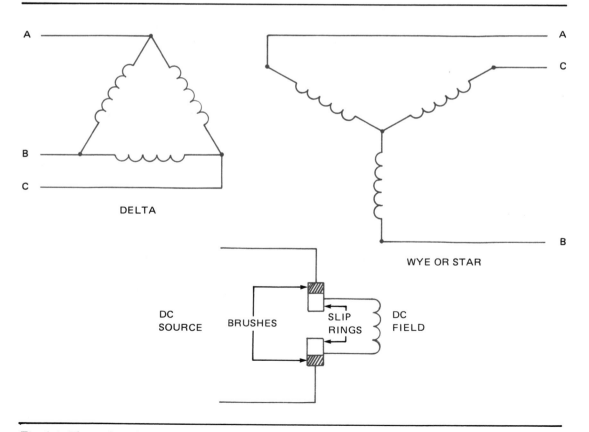

Fig. 1-2 Three-phase internal generator connections and a stationary armature with a rotating dc field

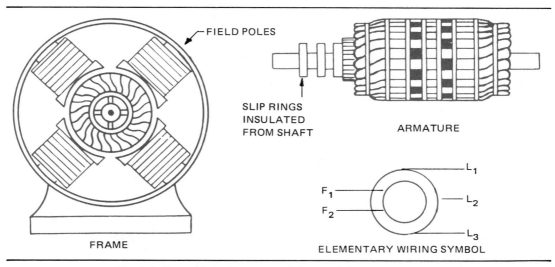

FIELD POLES

SLIP RINGS
INSULATED
FROM SHAFT

ARMATURE

FRAME

ELEMENTARY WIRING SYMBOL

Fig. 1-3 Parts of an alternator of the revolving-armature type

ALTERNATOR TYPES

Two principal types of synchronous alternators are: (1) the revolving-armature alternator and (2) the revolving-field alternator.

Figure 1-3 illustrates an alternator with a stationary field, a revolving armature, and the elementary wiring symbol for a three-phase alternator. The armature consists of the windings into which current is induced. The magnetic field for this type of alternator is established by a set of stationary field poles mounted on the periphery of the alternator frame. The field flux created by these poles is cut by conductors inserted in slots on the surface of the rotating armature. The armature conductors are arranged in a circuit which terminates in slip rings. Alternating current induced in the armature circuit is fed to the load circuit by brushes which make contact with the slip rings.

The revolving-armature alternator generally is used for low-power installations. The fact that the load current must be conducted from the machine through a sliding contact at the slip rings poses many design problems at higher values of load current. An alternator designed for limited use has semiconductor rectifier diodes installed on the exciter, thus eliminating the brushes and slip rings for the revolving field alternator.

FIELD EXCITATION

Direct current (dc) must be used in the field circuit of an alternator. As a result, all types of alternators must be supplied with field current from a dc source, except for small permanent magnet fields. The dc source may be a dc generator operated on the same shaft as the alternator. In this case, the dc generator is called an *exciter,* shown on the self-excited synchronous alternator in figure 1-4A. The circuit diagram for this alternator is shown in figure 1-4B. In installations where a number of alternators require excitation power, this power is supplied by a dc generator driven by a separate prime mover. The output terminals of this generator connect to a *dc exciter bus* from which other alternators receive their excitation power by means of brushes and slip rings for the revolving field alternator.

FIELD DISCHARGE CIRCUIT

A field discharge switch is used in the excitation circuit of an alternator. This switch eliminates the potential danger to personnel and equipment resulting from the high inductive voltage created when the field circuit is opened.

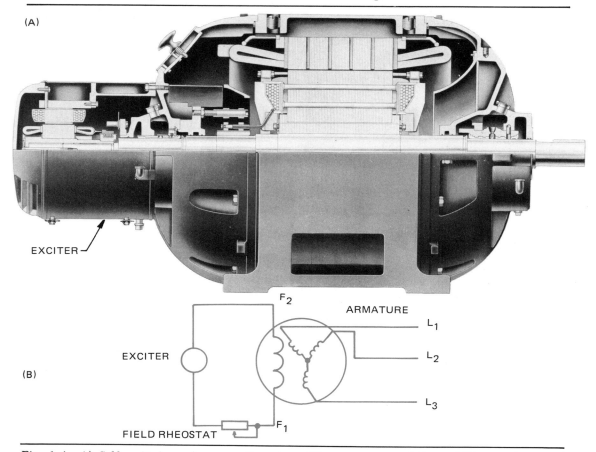

Fig. 1-4 A) **Self-excited synchronous alternator** *(Photo courtesy of General Electric Company)*
B) **Circuit diagram**

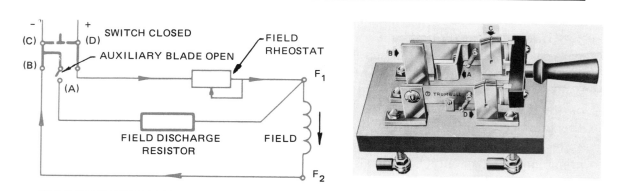

Fig. 1-5 Field discharge circuit   Fig. 1-6 Field discharge switch

Figure 1-5 illustrates the connections for the field circuit of a separately excited alternator. With the discharge switch closed, the field circuit is energized and the field discharge switch functions as a normal double-pole, single-throw switch.

The discharge switch shown in figure 1-6 has an auxiliary switch blade at A in addition to the normal blades at C and D (figure 1-5).

When it is desired to open the field circuit, the following actions must take place.

- Before the main switch contacts open, switch blade A meets contact B and thus provides a second path for the current through the field discharge resistor.
- When the main switch contacts C-D open (figure 1-7) high inductive voltage is created in the field coils by the collapsing magnetic field.
- This high voltage is dissipated by sending a current through the field discharge resistor.
- This procedure eliminates the possibility of damage to the insulation of the field windings as well as danger to anyone opening the circuit using a standard double-pole switch. This type of field circuit is used with all types of alternators.

FREQUENCY

The frequency of an alternator is a direct function of (a) the speed of rotation of the armature or the field and (b) the number of poles in the field circuit. Frequencies commonly used in the United States are fifty and sixty cycles per second or hertz (Hz). Power companies are particularly concerned with maintaining a constant frequency for their energy output since many devices depend on a constant value of frequency. This constant value is achieved by sensitive control of the prime mover driving the alternator and by voltage regulators.

If the number of field poles in a given alternator is known, then it is possible to determine the speed required to produce a desired frequency. One cycle of voltage is generated each time an armature conductor passes across two field poles of opposite magnetic polarity. The frequency in cycles per second or hertz is the number of pairs of poles passed by the conductor in a second. Since the speed of rotating machinery is

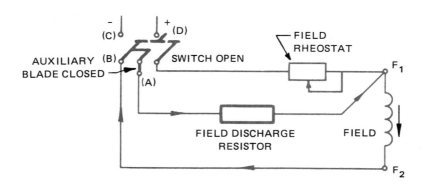

Fig. 1-7 Field discharge circuit

given in revolutions per minute (r/min), the speed in revolutions per second is obtained by dividing the speed (r/min) by 120. In a two-pole alternator the frequency is:

$$f = \text{poles} \times \frac{r/min}{120}$$

or

$$f = \frac{p \times S}{120}$$

Where f = frequency in hertz (formerly cycles per second)
    p = number of poles
    S = speed in revolutions per minute
  120 = alternations per second

The formula for frequency can be rearranged so that the speed required to give a desired frequency can be obtained.

$$S = \frac{120 \times f}{p}$$

If a two-pole alternator is to be operated at a frequency of 60 Hz, the correct speed is obtained from the formula $S = (120 \times f)/p$

$$S = \frac{120 \times 60}{2} = 3,600 \text{ r/min}$$

For a four-pole alternator operated at a frequency of 60 Hz, the required speed is:

$$S = \frac{120 \times 60}{4} = 1,800 \text{ r/min}$$

The two examples given illustrate the previous statement that the frequency of an alternator is a direct function of the speed of rotation and the number of poles in the alternator field circuit.

VOLTAGE CONTROL

The voltage output of an alternator increases as the speed of rotation increases or as the field excitation increases to the point of magnetic saturation of the field poles.

- For practical purposes, an alternator must be operated at a constant speed to maintain a fixed frequency.
- Thus, the only feasible method of controlling the voltage output is to vary the field excitation.

Field rheostats are used to vary the resistance of the total field circuit. This variation of resistance, in turn, changes the value of field current (figure 1-4B).

- A low value of field current results in less flux and less induced voltage at a given speed.
- A high field current results in greater field flux and a higher induced voltage at a given speed.
- The value of flux at which the field poles saturate determines the maximum voltage obtainable at a fixed speed and frequency.

ROTATING-FIELD ALTERNATORS

Rotating-field alternators are used extensively because of the ease with which a high-load current can be taken from the machine, reducing the number or eliminating the use of slip rings or sliding contacts. Thus, the use of rotating-field alternators results in a savings in initial cost and fewer maintenance requirements.

Stator Winding

Figure 1-8 illustrates the stator (stationary or nonmoving) windings of a rotating-field, three-phase alternator. The three-phase armature windings are embedded 120 degrees from one another in the slots of a laminated steel core which is clamped securely to the alternator frame. Output leads from the stator emerge from the bottom of the stator and connect directly to the load circuit. It can be seen that slip rings and brushes are not required in a stationary winding of this type. As a result, higher values of output voltage and current are possible. Standard values of voltage output for a rotating-field alternator are as high as 11,000 to 13,800 volts.

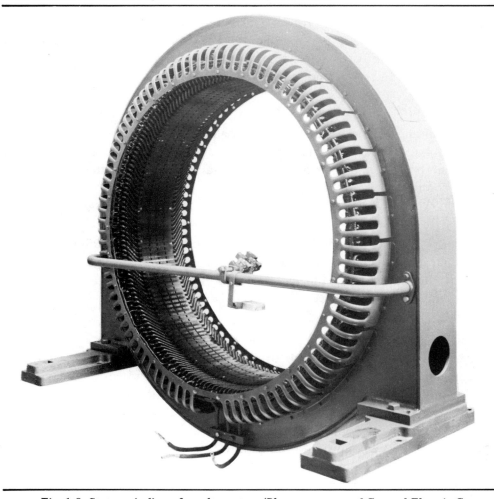

Fig. 1-8 Stator winding of an alternator *(Photo courtesy of General Electric Company)*

Rotating Field

The rotating portion of a rotating-field alternator consists of field poles mounted on a shaft which is driven by the prime mover. The magnetic flux established by the rotating field poles cuts across the conductors of the stator winding to produce the induced output voltage of the stator.

The following comparison can be made between the rotating-armature alternator and the rotating-field alternator. In the rotating-armature alternator, the armature conductors cut the flux established by stationary field poles. For the rotating-field alternator, the motionless conductors of the stator winding are cut by the flux established by rotating field poles. In each case an induced voltage is generated.

Figure 1-9 shows a salient field rotor for low-speed, three-phase alternators. For this type of rotor, the field poles protrude from the rotor support structure which is of steel construction and commonly consists of a hub, spokes, and rim. This support structure is called a *spider*. Each of the field poles is bolted to the spider. The field poles may be dovetailed to the spider in some alternators to provide a better support for the poles against the effects of centrifugal force.

Figure 1-10 shows a nonsalient rotor. This type of rotor has a smooth cylindrical surface. The field poles (usually two or four) do not protrude above this smooth surface. Nonsalient rotors are used to decrease windage losses on high-speed alternators, and to improve balance and reduce noise.

Power Supply for Rotor

The field windings of both salient and nonsalient rotors require dc power. Slip rings and brushes are used to feed the current to the windings at a potential of 100 to 250 volts dc. The brushes and rings are easily maintained because of the low values of field current encountered.

TERMINAL MARKINGS

A standard system of marking leads for field circuits has been established by the American Standards Association. The field leads for both alternators and generators are indicated by the markings F_1 and F_2. In addition, the F_1 lead always connects to the positive bus of the dc source. (See figures 1-5 and 1-7.)

ALTERNATOR REGULATION

Regardless of the type of generator or alternator used in a system, the terminal output voltage of the machine varies with any change in the load current. The impedance of the windings and the power factor of the load circuit both influence the regulation of an alternator. An increase in load current in a pure resistive load circuit causes a decrease in output voltage. A voltage drop of approximately 10 percent is common when going from a condition of no-load to full-load in a typical alternator.

For an inductive load, an increase in load current will cause a greater voltage drop than is obtained with a pure resistive load. A load with a low value of lagging power factor produces a large drop in output voltage.

Fig. 1-9 Alternator rotor, salient field type *(Photo courtesy of General Electric Company)*

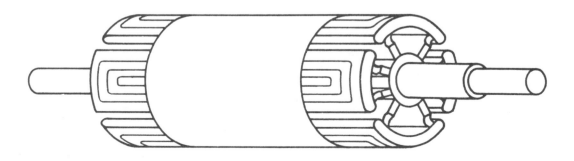

Fig. 1-10 Alternator rotor, nonsalient field type

A capacitive load circuit produces the opposite effect. In other words, the output voltage rises above the no-load value with an increase in load current and is high at a low value of leading power factor.

AUTOMATIC VOLTAGE CONTROL

Unlike dc generators, alternators cannot be compounded to alter the voltage-load characteristic. Moreover, output voltage variations are more likely to be severe because of changes in the load power factor. As a result, automatic voltage regulators generally are used with alternators.

Automatic voltage regulators change the alternator field current to compensate for any increase or decrease in the load current. A relay is used to increase or decrease the field resistance through contactors bridged across a field circuit resistor. As the ac line voltage falls, the relay bypasses sections of the field resistor to cause an increase in the flux and thus increase the induced voltage. An increase in the ac line voltage causes the relay to open contactors across the field resistor to decrease the field current, flux, and induced voltage. Power companies stabilize voltage by using a type of varying ratio transformer as a voltage regulator.

BRUSHLESS EXCITERS WITH SOLID-STATE VOLTAGE CONTROL

The permanent magnet generator (figure 1-11) supplies high-frequency ac power input to the voltage regulator. Voltage and reactive current feedback information is provided to the regulator from potential and current transformers. Using these feedback signals and a reference point established by setting the voltage adjusting rheostat, the voltage regulator (which has a transfer switch allowing the operator to select automatic regulator control or manual control) provides a controlled dc output. The dc is fed to the field of the rotating exciter; the three-phase, high-frequency ac output is then rectified by a full-wave bridge. This rectified signal is applied to the main generator field. Fully rated, parallel, solid-state diodes with indicating fuses are provided to

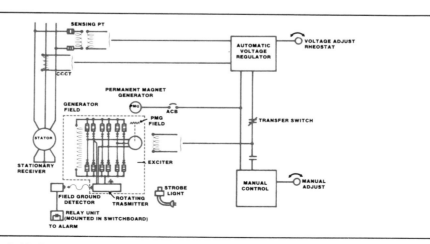

Fig. 1-11 Diagram of an exciter with permanent magnet generator *(Courtesy of Electric Machinery, Turbodyne Division, Dresser Industries, Inc.)*

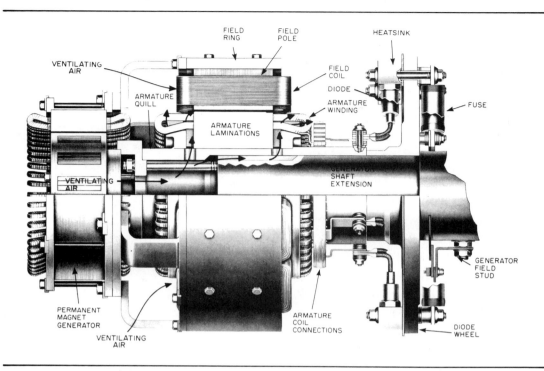

Fig. 1-12 Cutaway view of a brushless exciter showing the components *(Courtesy of Electric Machinery, Turbodyne Division, Dresser Industries, Inc.)*

Fig. 1-13 Rotating components of the brushless excitation system *(Courtesy of Electric Machinery, Turbodyne Division, Dresser Industries, Inc.)*

permit full load generation with a diode (rectifier) out of service. The use of a strobo-scope light permits the indicating fuses to be viewed during operation to determine if a diode has failed. Figure 1-12 shows a cutaway view of a brushless exciter. Figure 1-13 shows the rotating components of a brushless excitation system.

ACHIEVEMENT REVIEW

Select the correct answer for each of the following statements and place the cor-responding letter in the space provided.

1. The armature of an alternator   _____
    a.   is the revolving member.
    b.   is stationary.
    c.   is the frame.
    d.   consists of the windings into which the current is induced.

2. In alternators of the revolving-armature type,   _____
    a.   slip rings are required in the power output circuit.
    b.   slip rings are required in the field circuit.
    c.   slip rings are not required.
    d.   one slip ring is required.

3. In a protective field discharge circuit, the auxiliary blade of the
    field switch inserts the discharge resistor   _____
    a.   at the instant the field circuit opens.
    b.   immediately after the main blade loses contact.
    c.   immediately before the main blade loses contact.
    d.   immediately after the main blades make contact.

4. A field discharge circuit resistor   _____
    a.   is installed to stabilize line voltage.
    b.   is installed to stabilize line current.
    c.   improves regulation.
    d.   eliminates danger to people and equipment.

5. The frequency of the alternator output   _____
    a.   is directly proportional to its speed.
    b.   is inversely proportional to its speed.
    c.   depends upon its field strength.
    d.   is inversely proportional to the number of poles.

6. The speed of a six-pole, 60-Hz alternator is: _____
    a.   600 r/min   c.   1,800 r/min
    b.   1,200 r/min   d.   3,600 r/min

7. To deliver power at a frequency of 400 Hz, an eight-pole alternator
    must be driven at what speed?   _____
    a.   600 r/min   c.   6,000 r/min
    b.   3,600 r/min   d.   8,000 r/min

8. High-speed alternators are designed with   _____
   a. a revolving armature and a nonsalient rotor.
   b. a revolving armature and a salient rotor.
   c. revolving fields and a salient rotor.
   d. revolving fields and a nonsalient rotor.

9. Changing the driven speed of an alternator   _____
   a. changes the voltage magnitude up to field saturation.
   b. changes the frequency output.
   c. does not affect voltage or frequency.
   d. both a and b are correct.

10. The magnitude of the voltage output of an alternator is gen-
    erally controlled by   _____
    a. the speed of the prime mover.
    b. a field rheostat.
    c. variable resistance in the output lines.
    d. changing the power factor of the load.

11. Alternators use all but one of the following systems to obtain
    field excitation.   _____
    a. a separate dc power supply
    b. a self-excited ac field circuit
    c. a dc exciter on the same shaft as the alternator
    d. a rectifier to convert the output voltage for use in the
        field circuit

12. The greatest drop in output voltage results from taking full-
    load power from an alternator at a   _____
    a. unity power factor load.
    b. high power factor capacitive load.
    c. low power factor inductive load.
    d. medium power factor capacitive load.

13. Three-phase voltage is   _____
    a. three polyphase circuits.
    b. three distinct voltages.
    c. three delta connections.
    d. three wye connections.

14. The elementary wiring symbol for a three-phase alternator is   _____

ENGINE-DRIVEN GENERATING SETS

OBJECTIVES

After studying this unit, the student will be able to

- describe the purposes of engine-driven generating sets.
- list the advantages of using cogenerating sets.
- describe the operation of an automatic transfer switch.
- connect an automatic transfer switch.
- state National Electrical Code requirements.

ENGINE-DRIVEN GENERATING SETS

Diesel, gasoline, or natural gas engine-driven generators are most commonly used to provide another source of emergency or standby power when the normal utility power fails. Turbine power generator sets are also used in this application.

Sturdy, diesel-engine powered generators may lose some of their popularity in the remote, area-sites power systems. The use of hybrid systems using natural energy, such as the wind and the sun, are growing dramatically. So, although diesel generators will not become obsolete for remote site electrical power, a need will exist for a back-up source, as the generator changes its role from a primary energy source to part of a combined source.

Most engine-driven generator sets are rated from a few hundred watts to several hundred kilowatts, although units rated as high as 3,000 kW have been successfully applied. Multiple units, with some working in parallel, are becoming more commonly used to increase generating capacity. Controls may be manual, remote, or automatic, depending upon their application.

Transfer Switches

Switches are required to transfer, or reconnect, the load from a preferred or normal electric power supply to the emergency power supply from the generator set. This is done either manually or automatically. The manual method uses a double-throw switch, operated by hand, to transfer the load from the normal to the emergency power after the standby plant is already running. An automatic transfer switch (figure 2-1) usually starts and stops the standby power plant, and transfers the load by relays without requiring the attention of an operator.

An elementary diagram of a typical automatic transfer switch is shown in figure 2-2. (The figure does not include the engine starting controls and other controls.)

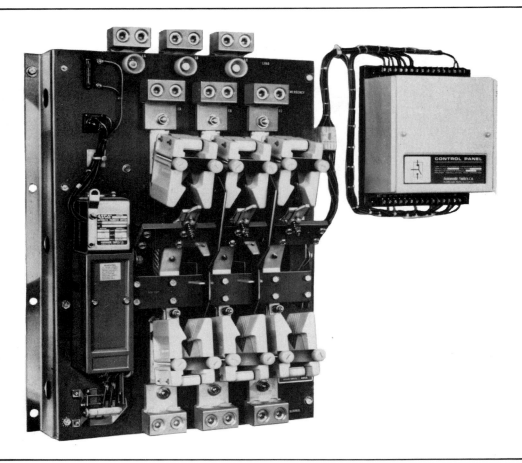

Fig. 2-1 A sophisticated, 600-ampere automatic transfer switch with accessory group control panel at right. Note the cable terminal lugs at top and bottom. *(Courtesy Automatic Switch Co.)*

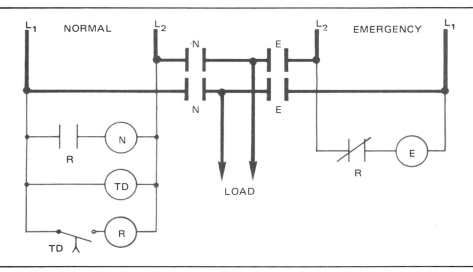

Fig. 2-2 Elementary diagram of an automatic transfer switch

When the normal supply on the left side is energized, current flows from L_1 through TD (time-delay coil) and back to L_2. After a predetermined setting of time delay in closing contact, relay R coil becomes energized. Contact R then closes, and energizes the N coil. Power contacts N then close, supplying the load from the normal or preferred source. When R coil is energized, it also opens the normally closed R contact interlock in the E coil emergency circuit. This safe action insures that each power supply operates independently of the other.

When the normal power fails, all coils on the left, or normal supply side, become deenergized. Relay contact R drops to its normally closed position in the E (emergency coil) circuit. Coil E is then energized, thereby closing the E power contacts feeding the load from an emergency electrical supply.

The time delay action helps to insure that the normal service does not supply the load intermittently with the emergency supply. In other words, the load will wait a preset time until the normal supply is firmly established before it is reconnected to it.

EMERGENCY SYSTEMS

Applicable National Electrical Code (NEC) and local code rules are considered when an on-site generator is selected. These differ, depending on whether the generating set is to function as a power source in a health care facility, such as a hospital, a standby power system, or as an emergency system.

Emergency generator systems generally are installed wherever great numbers of people gather, and where artificial lighting is required, such as in hotels, theatres, sports arenas, hospitals, and similar institutions. In addition to lighting, emergency systems supply loads which are essential to life and safety. Such installations include fire pumps, ventilation, refrigeration, and signaling systems when essential to maintain life. [Refer to *Article 700* of the National Electrical Code (NEC).]

Standby Power Generation Systems

Standby power generation systems include alternate power systems for applications such as heating, refrigeration, data processing, or communication systems where interruption of normal power would cause human discomfort or damage to the product in manufacture. (Refer to *Article 702* of the NEC.)

Health Care Facilities

Health care facilities are governed by several National Electrical Code rules concerning power sources, emergency systems, and essential electrical systems. In particular, refer to NEC *Sections 517-44C, 517-47,* and *517-65.*

Figure 2-3 shows a diesel-driven emergency power system consisting of four 450-kW electric generating sets. The system is electronically synchronized to deliver 1.8 million watts of emergency power for a hospital. Each unit can also be operated independently of the other units.

Fig. 2-3 1.8 million watt diesel-driven emergency power generating system
(Photo courtesy Onan Corporation, A Subsidiary of McGraw-Edison)

LEGALLY REQUIRED STANDBY SYSTEMS

NEC *Article 701* states that legally required standby power systems are those systems required by municipal, state, federal, or other codes or a government agency having jurisdiction. In the event of failure of the normal power source, these systems are intended to take over automatically. Legally required standby power systems are installed to serve such loads as communication systems, ventilation and smoke removal systems, sewage disposal, rescue and fire fighting equipment, among others. So, you see, there are installations that *must* be installed within the guidelines of the authority having jurisdiction.

COGENERATING PLANTS

Cogenerating plants are diesel-powered electric generators which are designed to recapture and use the waste heat both from their exhaust and cooling systems (figure 2-4).

Although cogenerating plants are not a new concept, they are now being used to combat the energy shortage and the rising prices charged by public utility companies for power generation. About a dozen of the nation's largest manufacturers of diesel

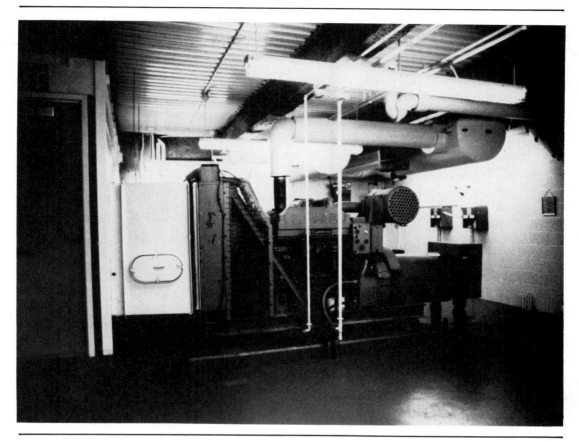

Fig. 2-4 A diesel-powered cogenerating plant *(Photo courtesy of Cummins Engine Corp.)*

engines have set out to provide competition for the foremost electric utility companies in the United States. As a result, these manufacturers have been concentrating on selling cogenerating plants.

Equipped with cogenerating plants, energy users need no longer rely on public utilities, due to the fact that not only can they make all of their own electricity, at a lower cost, but provide heating and cooling for their buildings, as well.

Various technical methods have been devised for using cogenerating plants. However, all of them capitalize on the fact that the generation of electricity wastes about twice as much energy in the form of heat as that amount of energy which can be generated as electricity. Steam heat, as a waste by-product of manufacturing processes, is now harnessed and used to turn steam turbine electric alternators. This electricity, when not needed, is sold to the public utility which services the plant.

The energy-saving application of cogeneration should result in greater demands for electrical work and, thus, more jobs for electricians. It also should create a particular need for power generator operators having the skills to install, operate, and maintain cogenerating equipment.

ACHIEVEMENT REVIEW

Select the correct answer for each of the following statements and place the corresponding letter in the space provided.

1. Engine-driven generating sets are used for _____
   a. emergency systems.
   c. cogenerating plants.
   b. standby power.
   d. all of these.

2. With an automatic transfer switch, as shown in figure 2-1, how does the emergency supply feed the load when power fails?
   a. TD energizes R.
   b. Normally open contact R opens.
   c. Normally closed contact R closes.
   d. Power contacts N close.

3. Generating capacities may be increased by using _____
   a. parallel multiple units.
   b. series multiple units.
   c. turbines.
   d. diesels.

4. Cogenerating sets are used _____
   a. to supply emergency power.
   b. to supply standby power.
   c. to conserve energy.
   d. in health care facilities.

5. Electrical capacity is gained with several small generating sets by _____
   a. paralleling machines on the line.
   b. reducing the load.
   c. placing machines on the line in series.
   d. none of these.

 # PARALLEL OPERATION OF THREE-PHASE ALTERNATORS

OBJECTIVES

After studying this unit, the student will be able to

- state the conditions which require that two alternators be paralleled.
- describe the use of synchronizing lamps in the three dark method and the two bright, one dark method of synchronizing alternators.
- demonstrate the procedure for paralleling two three-phase alternators.
- state the effect of changes in field excitation and speed on the division of load between paralleled alternators.
- describe "reverse power."

WHEN TO PARALLEL ALTERNATORS

Alternators are paralleled for the same reasons that make it necessary to parallel dc generators. Two alternators are paralleled whenever the power demand of the load circuit is greater than the power output of a single alternator.

When dc generators are paralleled, it is necessary to match the output voltage and electrical polarity of the machines with the voltage and polarity of the line. The same matching is required when alternators are paralleled. However, the matching of alternator polarity to that of the line presents problems not encountered when matching dc generator and line polarities. The output voltage of an alternator is continuously changing in both magnitude and polarity at a definite frequency. Thus, when two alternators are paralleled, not only must the rate of the rise and fall of voltage in both alternators be equal, but the rise and fall of voltage in one machine must be exactly in step with the rise and fall of voltage in the other machine. When two alternators are in step, they are said to be in *synchronism*. Alternators cannot be paralleled until their voltages, frequencies, and instantaneous polarities are exactly equal.

Figure 3-1 shows a comparison of the voltage curves of one of the phases of two three-phase generators operating independently but at different speeds. The voltage curves must be in synchronism before paralleling machines.

The output voltage of an alternator can be controlled by varying the strength of the direct current in the field circuit of the alternator. A field rheostat can be used to vary the dc current. Since the frequency of an alternator varies directly with speed changes, it is necessary to be able to control the speed of at least one alternator in an installation containing two machines.

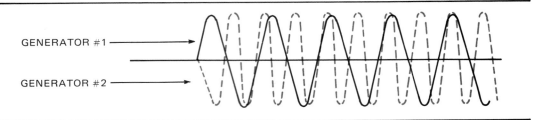

Fig. 3-1 Periodic time relationship of the out-of-phase voltages of two generators running at different speeds

ACHIEVING SYNCHRONIZATION

Three main steps are required to achieve synchronization.

1. The speed of one or both machines is varied so that *both* alternators generate output voltages at the desired frequency.
2. The ac voltage output of *both* machines is equalized using the field rheostats.
3. The frequency of one machine is increased or decreased by making slight changes in its speed until the frequencies of *both* machines are exactly equal in instantaneous polarity. The process of synchronization can be observed by an electrician using a synchroscope or connecting the alternators in parallel through a set of synchronizing lamps. The lamps indicate the exact instant at which the machines have like instantaneous electrical polarities. In addition, the lamps prevent the circulation of short-circuit currents between the alternators during the synchronizing process.

Frequencies, voltages and instantaneous ac polarities must be equal for synchronizing alternators. Two methods of synchronizing alternators are described as follows.

Three Dark Method

The following describes the method of synchronizing two alternators using the *three dark method.*

Figure 3-2 illustrates a circuit used to parallel two three-phase alternators. Alternator G_2 is connected to the load circuit. Alternator G_1 is to be paralleled with alternator G_2. Three lamps rated at double the output voltage to the load are connected between alternator G_1 and the load circuit as shown. When both machines are operating, one of two effects will be observed:

1. The three lamps will light and go out in unison at a rate which depends on the difference in frequency between the two alternators.
2. The three lamps will light and go out at a rate which depends on the difference in frequency between the two machines, but not in unison. In this case, the machines are not connected in the proper phase sequence and are said to be out of phase. To correct this, it is necessary to interchange any two leads to alternator G_1. The machines are not paralleled until all lamps light and go out in unison. The lamp method is shown for greater *simplicity* of operation.

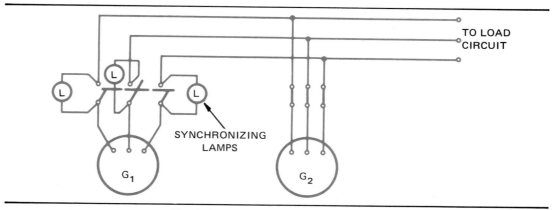

Fig. 3-2 Synchronization of alternators

By making slight adjustments in the speed of alternator G_1, the frequency of the machines can be equalized so that the synchronizing lamps will light and go out at the lowest possible rate. When the three lamps are out, the instantaneous electrical polarity of the three leads from G_1 is the same as that of G_2. At this instant, the voltage of G_1 is equal to and in phase with that of G_2. Now the paralleling switch can be closed so that both alternators supply power to the load. The two alternators are in synchronism, according to the three dark method.

The three dark method has certain disadvantages and is seldom used. A large voltage may be present across an incandescent lamp even though it is dark. As a result, it is possible to close the paralleling connection while there is still a large voltage and phase difference between the machines. For small capacity machines operating at low speed, the phase difference may not affect the operation of the machines. However, when large capacity units having low armature reactance operate at high speed, a considerable amount of damage may result if there is a large phase difference and an attempt is made to parallel the units.

Two Bright, One Dark Method

Another method of synchronizing alternators is the *two bright, one dark method.* In this method, any two connections from the synchronizing lamps are crossed after the alternators are connected and tested for the proper phase rotation. (The alternators are tested by the three dark method.) Figure 3-3A shows the connections for establishing the proper phase rotation by the three dark method. Figure 3-3B shows the lamp connections required to synchronize the alternator by the two bright, one dark method.

When the alternators are synchronized, lamps 1 and 2 are bright and lamp 3 is dark. Since two of the lamps are becoming brighter as one is dimming, it is easier to determine the moment when the paralleling switch can be closed. Furthermore, by observing the sequence of lamp brightness, it is possible to tell whether the speed of the alternator being synchronized is too slow or too fast.

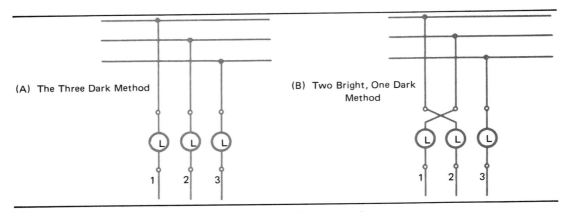

(A) The Three Dark Method   (B) Two Bright, One Dark
                                         Method

Fig. 3-3 Methods of synchronizing alternators

Synchroscope

A *synchroscope* is recommended for synchronizing two alternators since it shows very accurately the exact instant of synchronism (figure 3-4). The pointer rotates clockwise when an alternator is running fast and counterclockwise when an alternator is running slow. When the pointer is stationary, pointing upward, the alternators

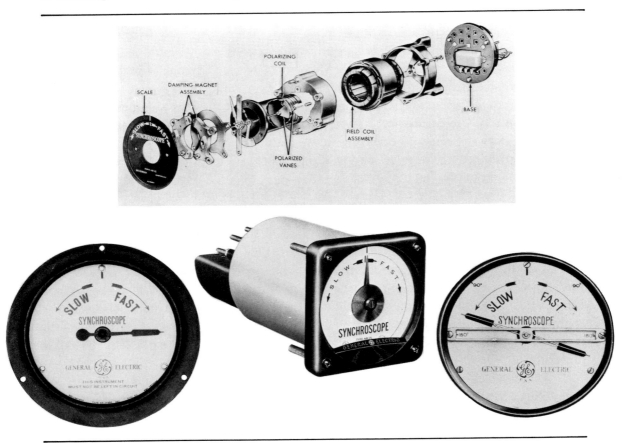

Fig. 3-4 A synchroscope *(Photo courtesy of General Electric Company)*

are synchronized. The synchroscope is connected across one phase only. For this reason it cannot be used safely until the alternators have been tested and connected together for the proper phase rotation. Synchronizing lamps or other means must be used to determine the phase rotation. In commercial applications, the alternator connections to a three-phase bus through a paralleling switch are permanent. This means that tests for phase rotation are not necessary. As a result, a synchroscope is the only instrument required to bring the machines into synchronism and thus parallel them.

Prime Movers

In industrial applications, alternators are driven by various types of prime movers such as steam turbines, water turbines, and internal combustion engines. For applications on ships, alternators often are driven by dc motors. Regardless of how alternators are driven, speed variation is a factor in paralleling the machines. Thus, the electrician should have a knowledge of speed governors and other speed regulating devices. This text, however, does not detail the operation of these mechanical devices.

PARALLELING ALTERNATORS

Since apprentices are likely to be required to parallel alternators driven by dc motors sometime in their instruction, the following steps outline the procedure for paralleling these machines. Figure 3-5 illustrates a typical circuit for paralleling two three-phase alternators.

Procedure

1. Set the field rheostat (R_2) of alternator G_2 to the maximum resistance position.
2. Knowing the number of field poles in alternator G_2, determine the speed required to generate the desired frequency.
3. Energize the prime mover to bring alternator G_2 up to the required speed.
4. Set switch S_3 to read the ac voltage across one phase of G_2. Adjust field rheostat R_2 until the output voltage is equal to the rated voltage of the load circuit.
5. Close the load switch and switch S_4 to feed the load circuit. Readjust the speed of the prime mover to maintain the predetermined speed required for the desired frequency.
6. Readjust R_2 to obtain the rated ac voltage of the load circuit.
7. Energize the prime mover to drive the second alternator, G_1. Adjust the speed of the alternator to the approximate value required to match the frequencies of the alternators.
8. Set switch S_3 to measure the ac voltage across one phase of G_1. Adjust field rheostat R_1 until the ac voltage is equal at either position of switch S_3. The voltage output of both alternators is now equal.
9. Phase Rotation
   With paralleling switch S_2 open, close switch S_1.
   The three sets of lamps across the terminals of the open switch will respond in one of two ways:

a. The three lamps will brighten and then dim in unison.

b. Two lamps will brighten in unison as the remaining lamp dims. Then the two bright lamps will dim as the dark lamp brightens.

10. If the lamps respond as in 9a, the alternators are connected for the proper phase rotation. The operator then may proceed to the next step in synchronizing the alternators.

11. If the lamps respond as in 9b, the alternators are not in the proper phase rotation. To correct the condition, interchange any two alternator leads at the terminals of

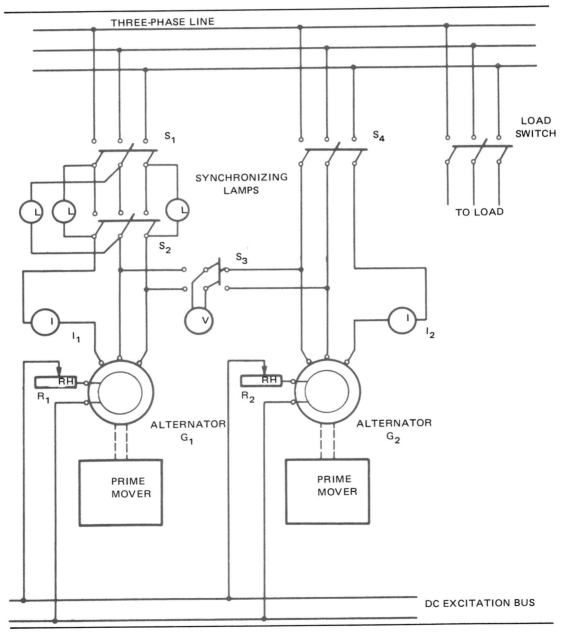

Fig. 3-5 Parallel operation of alternators

switch S_2. All three lamps should dim together and brighten together. No attempt to parallel the alternators should be made until the lamps respond in this manner.

12. The three lamp sets will flicker (dim and brighten) at a rate equal to the frequency difference between the two alternators. Adjust the speed control of prime mover M_1 to make the lamps flicker at the lowest possible rate.

13. Interchange two lamp set leads (not alternator leads) at the terminals of switch S_2 so that the alternators can be synchronized using the two bright, one dark method.

14. Again adjust the field rheostat of alternator G_1 until both alternators have the same output voltage as measured at either position of the voltmeter switch S_3.

15. With one hand on switch S_2, watch the lamps. Close the switch at the exact instant that two lamps are at their brightest and the other lamp is out. This operation shunts out the synchronizing lamps and parallels the alternators.

16. Ammeters I_1 and I_2 indicate the amount of load current carried by each alternator. If the load circuit has a unity power factor, then the sum of the ammeter readings should equal the reading of the ammeter in the load circuit.

17. Note that a change in the field excitation of either alternator does not appreciably change the amount of current supplied to the system. Such a change in field excitation does, however, affect the power factor of the specific alternator. The field rheostat of each machine should be adjusted to the highest power factor as indicated by the lowest value of current from the individual machine. Increasing or decreasing the mechanical power to either alternator will increase or decrease the load current of that machine. As a result, the division of the load between the alternators can be changed by slight changes in the alternator speed.

Speed vs. Load Characteristics

Two alternators operating in parallel must have the same frequency and the same terminal voltage. In addition, the prime movers of the parallel alternators must have similar drooping speed load characteristics. For steam-, diesel-, water-, or gas-driven prime movers, the speed load characteristic depends on adjustments of a mechanical speed control governor. These adjustments determine the division of load for two alternators operating in parallel. For this reason, the kilowatt load delivered by two alternators in parallel cannot be divided in any desired proportion by varying the dc field excitation of either machine.

Two alternators properly connected in parallel will operate in stable equilibrium. If one alternator attempts to pull out of synchronism, a current is created which circulates between both alternators. This current increases the speed of the lagging machine and retards the leading machine thus preventing the machines from pulling out of synchronism.

REVERSE POWER

If, for any reason, one machine is allowed to slow to a point where the other machine is taking all the electrical load, the zero load generator then goes to a negative value or "reverse power." This generator has now become a motor. This situation is of particular concern where the machine's protective scheme has not been designed to operate properly in the motoring situation. In such conditions reverse-current relays are usually employed to trip the generator on detection of reverse power flow.

ACHIEVEMENT REVIEW

A.   Select the correct answer for each of the following statements and place the corresponding letter in the space provided.

1. Two alternators are paralleled   _____
    a.   so that one is not overworked.
    b.   because of a rising load demand.
    c.   to ease the workload.
    d.   because of the declining load demand.

2. To parallel alternators, it is necessary to match   _____
    a.   voltages.
    b.   frequencies.
    c.   voltages and frequencies.
    d.   voltages, frequencies, and instantaneous polarities.

3. The output voltage of an alternator is controlled by   _____
    a.   adjusting the prime mover.
    b.   adjusting the direct current of the field circuit.
    c.   synchronizing lamps.
    d.   a synchroscope.

4. Alternators should not be paralleled unless the synchronizing
    lamps are lighting and dimming   _____
    a.   in rotation.
    b.   in reverse rotation.
    c.   in unison.
    d.   alternately.

5. Three lights flashing rapidly in unison while paralleling alternators
    means that   _____
    a.   the machines are not polarized.
    b.   the phase sequences are wrong.
    c.   the paralleling switch should be closed.
    d.   the frequencies differ by a large amount.

6. The three dark method of synchronizing alternators has the disadvantage that   _____
    a.   the lamps may burn out.
    b.   an undetected voltage may be present at the lamps.
    c.   the light is more difficult to see.
    d.   an undetected current may be present through the lamps.

7. The most reliable method of synchronizing alternators is to use _____
   a. a synchroscope.
   b. the three dark method.
   c. the three light method.
   d. the two bright, one dark method.

8. If a synchroscope is rotating clockwise, the _____
   a. alternators are ready to parallel.
   b. alternator being synchronized is too slow.
   c. alternator being synchronized is too fast.
   d. machines have not been polarized.

9. When the pointer of a synchroscope is stationary and points upward during the paralleling operation, the _____
   a. alternators are in synchronism.
   b. alternators are not in synchronism.
   c. incoming alternator frequency is too slow.
   d. incoming alternator frequency is too fast.

10. The division of load between alternators operating in parallel is
    accomplished by changing the _____
    a. field excitation.
    b. speed of the prime movers.
    c. power factor of the load.
    d. machine characteristics.

B. Insert the word or phrase to complete each of the following statements.

1. To operate satisfactorily in parallel, two alternators must have the same _____
   _____ , the same frequency, and the same _____ .

2. Two alternators are to be connected in parallel. The best instrument to use for
   synchronizing them is a(an) _____ .

3. An alternator is connected to a live three-phase bus. Using the three dark method,
   a lamp is connected in series with each lead. The lamps brighten and dim in unison. This proves that the alternators have the proper _____
   rotation.

4. In question 3, the switch shorting the three series lamps should be closed at the
   instant the lamps are _____ .

5. Two 208-volt alternators are to be paralleled. The synchronizing lamps should be
   rated at _____ .

6. The output voltage of alternators operating in parallel is equalized by adjusting
   their _____ .

7. The load on an alternator operating in parallel with another alternator may be increased by decreasing the spring tension of its speed _____ .

8. The division of load between two alternators operating in parallel can be changed by adjusting the _____ .

9. Two alternators, A and B, are being synchronized for parallel operation. Alternator A is operating at a frequency of 60 hertz. The synchronizing lamps are flickering twice a second. The frequency of alternator B is _____ hertz or _____ hertz.

10. Synchronizing lamps and a synchroscope are being used to parallel two alternators. Just before the moment the alternators are paralleled, there is no visible light from the lamps but the synchroscope is rotating slowly. In this case, the _____ method should be used to indicate when the paralleling switches should be thrown because _____ _____ .

WIRING FOR ALTERNATORS

OBJECTIVES

After studying this unit, the student will be able to

- describe the connections for and the resulting operation of the direct-current field excitation circuit for an alternator.
- describe the connections for and the resulting operation of the alternator output circuit for an alternator.
- describe the connections for and the resulting operation of the instrument circuits for an alternator.

This unit is concerned with the control panel and equipment for a three-phase, 2,400-volt alternator. The circuits and connections covered in detail are the direct-current field circuit and all control equipment; the alternating-current, three-phase output circuit with associated switchgear; and the connections for the instruments and instrument transformers used in a common installation.

DIRECT-CURRENT CIRCUIT FOR FIELD EXCITATION

The direct-current circuit requires dc bus bars, a field switch with a field discharge resistor, a dc ammeter with an external shunt, and a field rheostat. The field rheostat may be mounted on the back of the control panel with the insulated handle extending through to the front of the panel. If the field rheostat is very large, however, it cannot be mounted on the back of the switchboard; it can be mounted either near the ceiling above or in a room directly below the switchboard. In situations where large rheostats are located at a distance from the control panel, a chain and sprocket arrangement is used to connect the rheostat to the rheostat handle mounted on the control panel. As a result, the rheostat can be adjusted at the control panel.

Figure 4-1 illustrates the connections required for the separately excited field circuit of an alternator. Note that when the field discharge switch is open, the auxiliary blade closes to complete a path through the field discharge resistor. Thus, any inductive voltage in the alternator field is discharged through the field discharge resistor to prevent damage. **The field rheostat is connected so that it is not in the discharge circuit.**

ALTERNATOR OUTPUT CIRCUIT

The alternator in the installation described in this unit is rated at 2,400 volts, three phase. The three-phase, 2,400-volt output of the alternator is fed to the switchboard

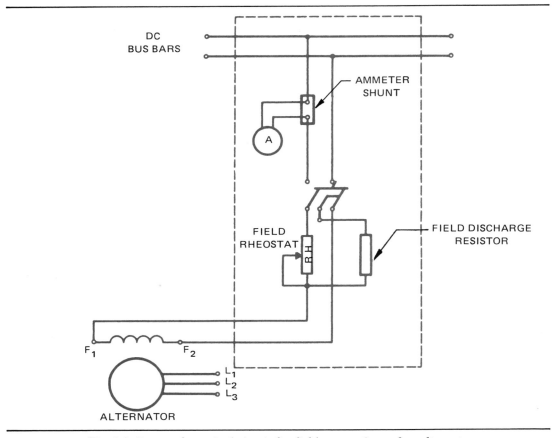

Fig. 4-1 Separately excited circuit for field connections of an alternator

through a three-wire, high-voltage lead cable in galvanized rigid conduit. The three conductors are fed through an oil-type circuit breaker, current transformers, and disconnect switches to the three-phase bus bars. An oil-type circuit breaker (switch) is used because of the relatively high voltage of the alternator. As the contacts of this switch open, any arc is immediately quenched in insulating oil.

Figure 4-2 illustrates an electrically-operated oil switch (circuit breaker). Note that each of the three sets of contactors is mounted in a separate cell or tank which is filled with an insulating oil. The three sets of contactors thus open and close in oil. The figure also shows a contactor assembly for one pole of a three-pole oil switch. Note the closing coil and the trip coils. The closing coil is relatively large and has a very fast positive action; the trip coil is smaller in size. The trip coil actuates a trip latch which causes the oil switch contactors to open.

The control circuit for the oil switch in a majority of alternator installations is connected to a dc source such as a bank of batteries. If there is a complete failure of the ac power, the oil switch can still be operated from the dc source, as is true of other emergency circuits.

A small switch handle located on the switchboard is used to adjust the control circuit. Two indicating lamps also are mounted on the switchboard. One of the indicating

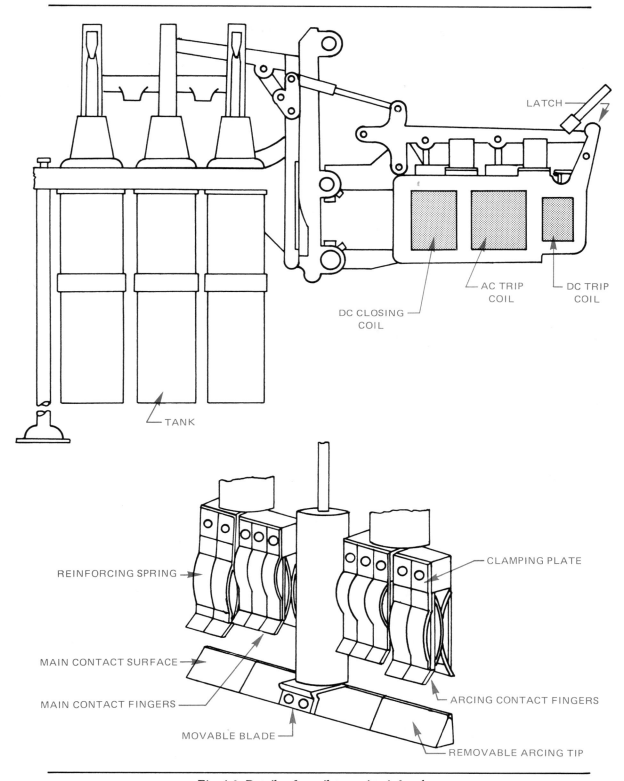

Fig. 4-2 Details of an oil-type circuit breaker

lamps is green and is on when the oil switch is open. The second indicating lamp is red and is on when the oil switch is closed. The red lamp normally is located directly above the control switch handle and the green lamp is located below the switch handle.

Figure 4-3 is the schematic connection diagram of the control circuit for the oil switch. When the oil switch is in the open or off position, the green pilot lamp is on. Note that there is a path from the positive side of the line through the current-limiting resistor, through the green indicating lamp, and through the normally closed M contacts to the negative side of the line.

When the on (start) button is pressed, a circuit is established from the positive side of the line to the control relay and then to the negative side of the line. The control relay is energized and closes its contacts to establish a path through the main closing

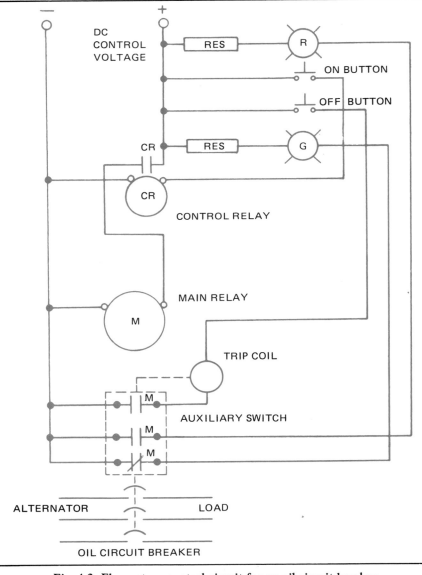

Fig, 4-3 **Elementary control circuit for an oil circuit breaker**

coil. The three main sets of oil switch contacts also close at this time. When the main closing relay is energized, the normally closed M contacts open. In addition, the green pilot lamp circuit opens and the two normally open M contacts close. The red indicating lamp is now on. When the on button is released, the oil switch remains in the on position due to the fact that it is secured by a latch mechanism.

When the off button is pressed, the trip coil is energized to trip the latch mechanism. The oil switch contacts thus open to the off position. As a result, the red indicating lamp goes out and the green indicating lamp lights.

The control handle and indicating lamps for an oil switch generally are mounted on the switchboard. The oil switch itself, however, is usually, but not always, located in a separate fire-proof room or vault below the switchboard room.

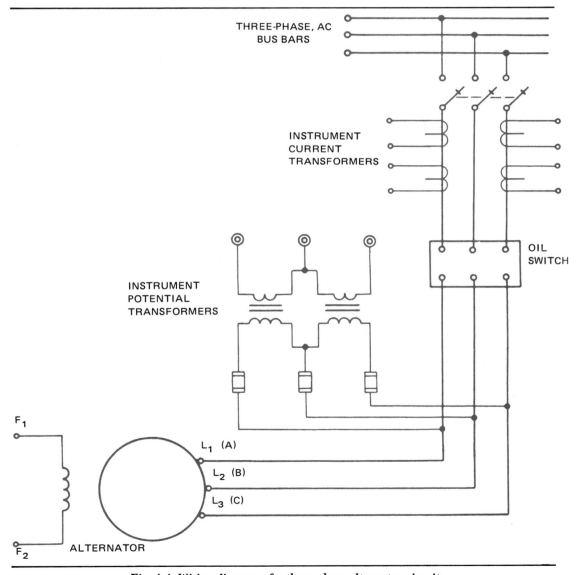

Fig. 4-4 Wiring diagram of a three-phase alternator circuit

Current transformers are used to step down the current in the output leads of the alternator to a value which can be used in instrument circuits. Step-down current transformers also insulate the low-voltage instrument circuit from the high-voltage primary circuit. The secondary current rating of a current transformer is 5 amperes. The current rating of the primary winding of the transformer must be high enough to handle the maximum current delivered by the alternator.

The alternator output leads feed from the current transformers to disconnect switches and then to the three-phase bus bars. A disconnect switch is a form of knife switch which is opened with a switch stick while exposed to air. The disconnect switches are operated *only* after the alternator oil switch is opened. The operator must wear rubber gloves when using an approved switch stick to open the disconnect switches. *Never open disconnect switches under load;* this is the purpose of the oil switch. It is designed to interrupt the arc without damage.

In most alternator installations, the three-phase bus bars are energized constantly. Since the disconnect switches disconnect the oil switch and the alternator from the bus bars, the alternator can be shut down and the disconnect switches opened to permit maintenance work on the oil switch under safe conditions. When the alternator requires maintenance or repair work, the disconnect switches are pulled to the off position even though the oil switch is open. The reason for this precaution is that the insulating oil in the oil switch may have become carbonized. The carbonized oil can act as a partial conductor resulting in a feedback from the live 2,400-volt bus bars through the oil switch and carbonized oil to the alternator terminals. Remember, then, that the disconnect switches and the oil switch must be *open* when any maintenance or repair work is to be done on ac generators. The generators should also be shut down.

Figure 4-4 is a wiring diagram of typical alternator connections to the three-phase bus bars.

The three bus bars for the ac output of the alternator are mounted on insulators, because the bus bars have a potential difference of 2,400 volts between them. It is important that the proper air gap be maintained between the three bus bars and that adequate clearance be provided between the bus bars and the ceiling and side walls of the room. Barriers shall be placed in all service switchboards to isolate the service bus bars and terminals from the remainder of the switchboard.

The National Electrical Code (*Article 384*) provides guidelines for switchboard and panelboard installations.

Figure 4-5 shows a typical control panel arrangement. The field rheostat handle, the control handle for the oil switch, and the field discharge resistor switch are all located in the same dead front panel section. The two indicating lamps for the oil switch are located above and below the oil switch control on the same panel section. Note that the field rheostat is located here in a room below the switchboard, but is operated remotely from the panel in this example.

The top section of the control panelboard contains several measuring instruments and their corresponding switches. The side view of the switchboard on the right of figure 4-5 shows the arrangement of the bus bars, disconnect switches, and instrument transformers.

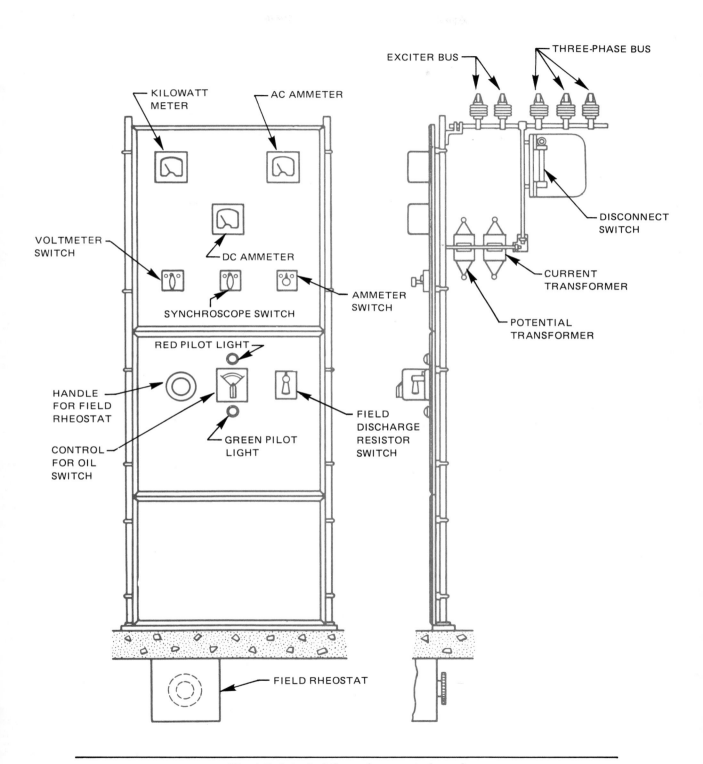

Fig. 4-5 A typical three-phase control panel

INSTRUMENT CIRCUITS

The voltage to the potential coils of instruments mounted on the switchboard rarely exceeds 120 to 125 volts. The voltage coils of wattmeters, watthour meters and voltmeters usually are designed for a maximum voltage of 150 volts. Since the three-phase output of the alternator is 2,400 volts, two instrument potential transformers connected in open delta are required to step down the voltage to 120 volts, three phase. The potential transformers are small in size since the load on the low-voltage secondary is very small. Each potential transformer is rated at 100 to 200 volt-amperes (VA). For the installation shown in figure 4-5, the load on the secondary of the transformer consists of the potential coils of the kilowatt meter and the voltmeter. The instrument potential transformers are rated at 2,400 volts on the high-voltage side and 120 volts

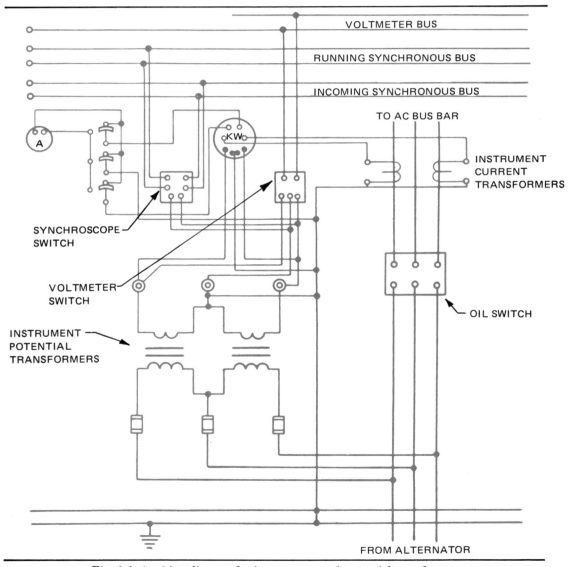

Fig. 4-6 A wiring diagram for instruments and potential transformers

on the low-voltage side. The low voltage at the instruments allows maintenance electricians to work more safely when making adjustments and repairs to the instruments.

The current coils of the measuring instruments mounted on switchboards usually are rated at a maximum current capacity of 5 amperes. In figure 4-5, each of the two current coils of the three-phase kilowatt meter is connected in series with an ammeter selector switch across the secondary of the proper current transformer.

It is unsafe to open the secondary circuit of a current transformer when there is a current flow in the primary circuit. (See ELECTRICITY 3, *Instrument Transformers*.)

Figure 4-6 is a wiring diagram for most of the instruments and instrument transformers shown in figure 4-5. A selector switch is used to insert the ac ammeter in the secondary circuit of either of the two current transformers. The current in the secondary

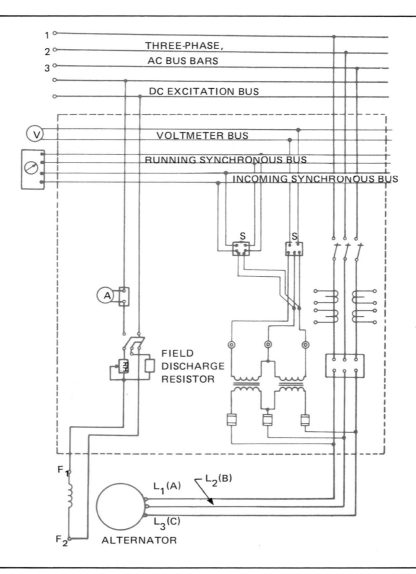

Fig. 4-7 Circuit connections for voltmeter and synchroscope

circuits is never in excess of 5 amperes. Therefore, either No. 14 or No. 12 AWG wire is used on the rear of the switchboard.

For a majority of permanent switchboard installations, the scale readings on the instruments are graduated to include the voltage and current transformer multipliers. This means that any error made by the switchboard operator in applying instrument multipliers is automatically eliminated.

Two instruments not shown in figure 4-5 or in the wiring diagram in figure 4-6 are the voltmeter and the synchroscope. In typical installations, there may be several alternators operating in parallel. Each alternator has a separate panel and these panels are mounted next to one another to make up a complete switchboard. One voltmeter and one synchro-scope are then mounted on a movable arm located at the end of the switchboard. The position of this arm can be adjusted so that the voltmeter and synchroscope are visible from any one of the generator control panels. A voltmeter switch located on each generator panel gives the operator a means of connecting the voltmeter to measure the voltage output of any alternator. In addition, special synchronizing switches permit the use of one synchroscope to synchronize any one of several alternators to the three-phase system.

Figure 4-7 shows the circuit connections for the voltmeter and synchroscope. Figure 4-6 indicates that the voltmeter switch has three positions. The voltmeter can be connected across any one of the three voltages of an alternator. If the voltage of a second alternator must be measured, the voltmeter switch is turned to the off position. The switch handle or key is then removed and inserted in the voltmeter switch of the second ac generator. Again, the switch may be turned to any one of the three voltage positions. Thus, one voltmeter can be used to measure the three voltages of each of several ac generators controlled through the switchboard.

A synchroscope switch is mounted on each alternator panel. When the switch handle is turned to the incoming position, the synchroscope is connected to the secondary volt-age of one phase of an alternator being synchronized with the ac system. The synchro-scope switch of a second alternator, which is already paralleled with the three-phase system, is connected to the run position. Thus, one coil winding of the synchroscope is energized from the running bus bars. The other winding of the synchroscope is energized from the incoming bus bars. With these connections, the synchroscope will indicate the extent the incoming machine is out of phase. When the incoming alterna-tor is in phase with the three-phase system, and the alternator voltage is equal to that of the bus bars, the control switch can be turned to the on position. As a result, the oil switch contactors close and the alternator is paralleled with the bus bars. The oil circuit breaker is used to connect and disconnect the alternator when it is running under load. This insures safe operation and prolongs switch contact life.

ACHIEVEMENT REVIEW

1. What is the purpose of disconnect switches in an ac generator installation?

    _____

    _____

    _____

2. Why is direct current used on the control circuits of oil switches used in alternator installations? _____

_____

_____

_____

3. Why is an oil switch normally used to interrupt the power output of an alternator?

_____

_____

_____

_____

4. Why are instrument transformers used for the instrument circuits of ac generator installations? _____

_____

_____

SUMMARY REVIEW
OF UNITS 1-4

OBJECTIVE

- To give the student an opportunity to evaluate the knowledge and understanding acquired in the study of the previous four units.

A. Insert the word or phrase to complete each of the following statements.

1. The main three-phase leads from a high-voltage alternator usually feed the main bus bars through a switch, the contacts of which are covered by _____ .

2. To minimize the danger to personnel working on the maintenance of high-voltage, three-phase alternators, a _____ switch is used in the main three-phase output leads.

3. In the event of failure of the dc supply used to control the main switch of a three-phase, high-voltage alternator, a separate dc source consisting of _____ is used.

4. An indicating lamp is used to indicate that the main switch is closed on a three-phase, high-voltage alternator. This lamp is colored _____ .

5. A _____ indicating lamp is used to indicate that the main line switch is open.

6. Current is measured in the three-phase leads of a high-voltage alternator by ammeters connected in the output leads of the alternator through the _____ .

7. The disconnect switch in the main line of an alternator is opened and closed by an operator using rubber gloves and a _____ .

8. A voltage feedback from the main bus to the output terminals of the alternator can occur through an open oil switch as the oil becomes _____ .

9. Voltage measurements are made on high-voltage alternators with voltmeters connected to the line through _____ .

10. The regulation of an alternator is influenced by the impedance of its windings and the _____ of the load circuit.

11. The speed of the prime mover driving an alternator determines the _____ and _____ of the output.

12. The normal voltage regulation of an alternator is least affected by a load with a slightly _____ power factor.

13. The output voltage of alternators is maintained through the use of voltage _____ .

14. When paralleling two alternators, the procedure used to bring both machines to the same exact phase relationship is called _____ .

15. The paralleling of alternators without a synchroscope is best accomplished with synchronizing lamps using the _____ method.

16. The voltage output of an alternator is controlled by adjusting the _____ circuit resistance.

17. In a revolving-field alternator, slip rings are used to conduct current to the _____ _____ circuit.

18. A dc generator mounted on the same shaft as the alternator is referred to as the _____ .

19. At a fixed speed of rotation, the frequency of the output voltage depends on the number of _____ .

20. Alternator field windings are marked with the letters _____ and _____ .

21. The extent to which voltage output decreases with increases in load current is referred to as voltage _____ .

22. The three-phase windings and the laminated core of a three-phase alternator of the rotating-field type are known as the _____ .

23. An alternator with four field poles is to generate power at 60 hertz. For this frequency the speed must be _____ r/min.

24. An increase in the field current of an alternator increases its output voltage to an extent determined by field _____ .

25. The regulation of an alternator is poorest when the load circuit has a low, _____ _____ power factor.

B. Select the correct answer for each of the following statements and place the corresponding letter in the space provided.

1. When *removing* the load from an alternator, the _____
   a. oil switch should be opened.
   b. disconnect switch should be opened first.
   c. machine should be slowed.
   d. machine should be stopped.

2. Current to the ammeters on an alternator installation switchboard is never in excess of _____
   a. 100 amperes.
   b. 2,400 volts.
   c. 50 amperes.
   d. 5 amperes.

3. A main disconnect switch is used to _____
    a. remove the load from the alternator.
    b. disconnect the oil switch and alternator from the energized bus bar.
    c. energize the bus bar.
    d. energize the oil switch and alternator.

4. Alternator installation switchboard voltmeters are connected to _____
    a. potential transformers.
    b. current transformers.
    c. the hot bus bar.
    d. the oil switch.

5. An oil switch is used to _____
    a. remove the disconnect switch from the line.
    b. energize the alternator.
    c. interrupt high voltages and currents.
    d. lubricate the disconnect switch.

6. When using a synchroscope to parallel alternators, the switches are closed when the indicator is _____
    a. revolving clockwise.
    b. revolving counterclockwise.
    c. pointing straight up.
    d. oscillating.

7. Adjusting the speed of the prime mover of an alternator causes a change primarily in the _____
    a. voltage.
    b. frequency.
    c. phase polarity.
    d. phase poles.

8. The voltage output of an alternator should be increased or decreased by _____
    a. adjusting the field rheostat.
    b. adjusting the speed.
    c. changing the number of poles.
    d. changing the capacities.

9. In an automatic transfer switch, the purpose of the time delay relay is to _____
    a. allow the engine-driven generator to pick up speed.
    b. permit the load to increase.
    c. delay the normal power supply until it is firmly established.
    d. delay the emergency power supply until it is firmly established.

THREE-PHASE, SQUIRREL-CAGE INDUCTION MOTOR

OBJECTIVES

After studying this unit, the student will be able to

- describe the construction of a three-phase, squirrel-cage motor, listing the main components of this type of motor.

- identify the following items and explain their importance to the operation of a three-phase, squirrel-cage induction motor: rotating stator field, synchronous speed, rotor induced voltages, speed regulation, percent slip, torque, starting current, no-load power factor, full-load power factor, reverse rotation, and speed control.

- calculate motor speed and percent slip.

- reverse a squirrel cage motor.

- describe why a motor draws more current when loaded.

- draw diagrams showing the dual voltage connections for 230/460-volt motor operation.

- explain motor nameplate information.

OPERATING CHARACTERISTICS

The three-phase, squirrel-cage induction motor is relatively small in physical size for a given horsepower rating when compared with other types of motors. The squirrel-cage induction motor has very good speed regulation under varying load conditions. Because of its rugged construction and reliable operation, the three-phase, squirrel-cage induction motor is widely used for many industrial applications (figure 6-1).

CONSTRUCTION DETAILS

The three-phase, squirrel-cage induction motor normally consists of a stator, a rotor, and two end shields housing the bearings that support the rotor shaft.

A minimum of maintenance is required with this type of motor because

- the rotor has no windings to become shorted.

- there are no commutator or slip rings to service (compared to the dc motor).

- there are no brushes to replace.

The motor frame is usually made of cast steel. The stator core is pressed directly into the frame. The two end shields housing the bearings are bolted to the cast steel frame. The bearings which support the rotor shaft are either sleeve bearings or ball

Fig. 6-1 Industrial application of a squirrel-cage motor   *(Courtesy General Electric Company)*

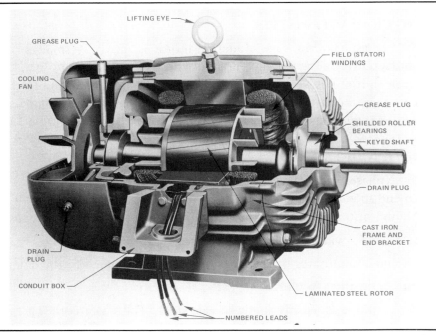

Fig. 6-2   Cutaway view of the construction and features of a typical three-phase explosion-proof motor *(Photo courtesy of Marathon Electric Manufacturing Corp.)*

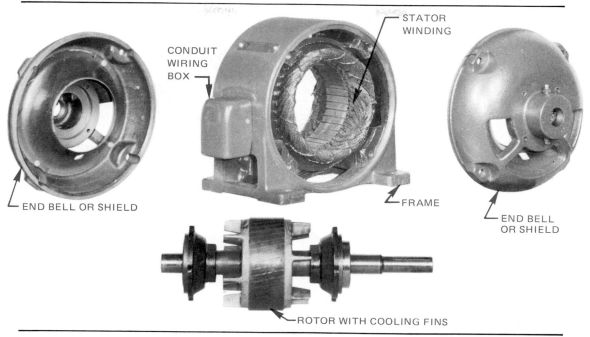

Fig. 6-3 Main components of a squirrel-cage induction motor *(Courtesy General Electric Company)*

bearings. Figure 6-2 is a cutaway view of an assembled motor. Figure 6-3 illustrates the main parts of a three-phase, squirrel-cage induction motor.

Stator

A typical stator contains a three-phase winding mounted in the slots of a laminated steel core (figure 6-4). The winding itself consists of formed coils of wire connected so that there are three single-phase windings spaced 120 electrical degrees apart. The

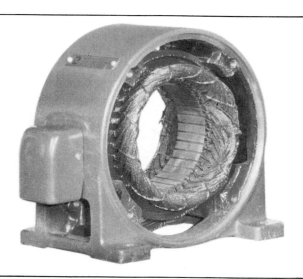

Fig. 6-4 Wound stator of an induction motor *(Courtesy General Electric Company)*

Fig. 6-5   Cage rotor for an induction motor   *(Courtesy General Electric Company)*

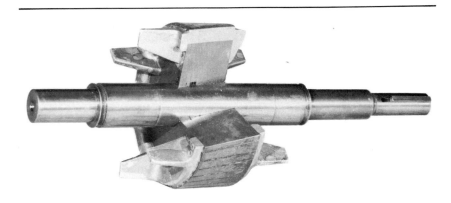

Fig. 6-6   Cutaway view of a cage rotor   *(Courtesy General Electric Company)*

Fig. 6-7   Squirrel-cage form for an induction motor   *(Courtesy General Electric Company)*

Fig. 6-8   A ball-bearing, open, squirrel-cage poly-phase induction motor   *(Photo courtesy of Newman Electric Motors)*

three separate single-phase windings are then connected, usually internally, in either wye or delta. Three or nine leads from the three-phase stator windings are brought out to a terminal box mounted on the frame of the motor for single- or dual-voltage connections (see unit 7).

Rotor

The revolving part of the motor consists of steel punchings or laminations arranged in a cylindrical core (figures 6-5 to 6-7). Copper or aluminum bars are mounted near the surface of the rotor. The bars are brazed or welded to two copper end rings. In some small squirrel-cage induction motors, the bars and end rings are cast in one piece from aluminum.

Figure 6-5 shows such a rotor. Note that fins are cast into the rotor to circulate air and cool the motor while it is running. Note also that the rotor bars between the rings are at an angle to the faces of the rings. Because of this design, the running motor will be quieter and smoother in operation. A keyway is visible on the left end of the shaft. A pulley or load shaft coupling can be secured using this keyway.

Shaft Bearings

Typical sleeve bearings are shown in figures 6-9 and 6-10. The inside walls of the sleeve bearings are made of a babbitt metal which provides a smooth, polished, and long wearing surface for the rotor shaft. A large oversized oil ring fits loosely around the rotor shaft and extends down into the oil reservoir. This ring picks up and slings oil on the rotating shaft and bearing surfaces. Two oil rings are shown in figure 6-11. This lubricating oil film minimizes friction losses. An oil inspection cup on the side of each

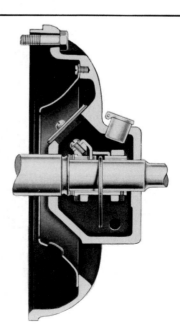

**Fig. 6-9 Sleeve-bearing end shield
for an open polyphase motor
*(Courtesy General Electric Company)***

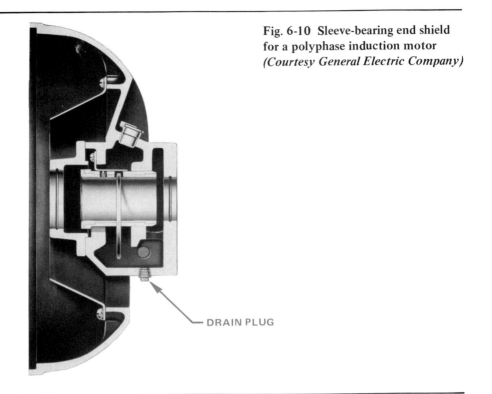

Fig. 6-10 Sleeve-bearing end shield for a polyphase induction motor *(Courtesy General Electric Company)*

DRAIN PLUG

OIL RINGS

Fig. 6-11 Partially assembled sleeve bearing for a totally enclosed, 1,250-hp motor *(Photo courtesy of Siemens-Allis)*

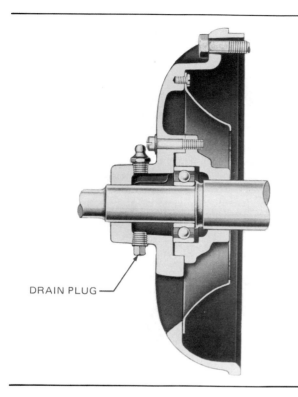

DRAIN PLUG

Fig. 6-12 Ball bearing end shield for an open polyphase motor *(Photo courtesy of General Electric Company)*

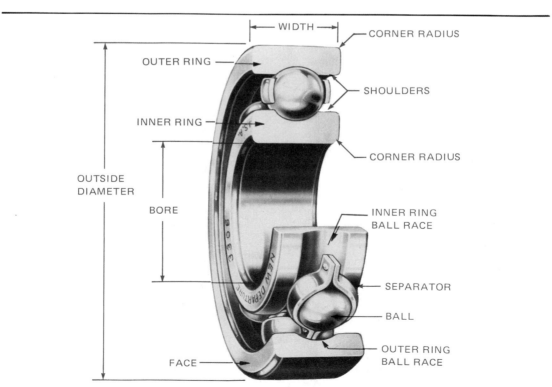

WIDTH

CORNER RADIUS

OUTER RING

SHOULDERS

INNER RING

CORNER RADIUS

OUTSIDE DIAMETER

BORE

INNER RING BALL RACE

SEPARATOR

BALL

OUTER RING BALL RACE

FACE

Fig. 6-13 Cutaway section of a single-row ball bearing *(Photo courtesy of New Departure Division, General Motors Corporation)*

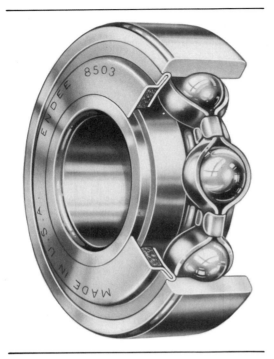 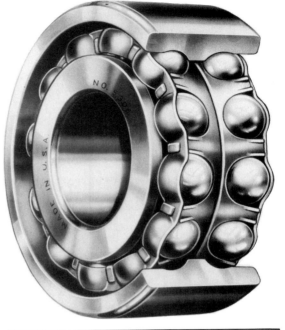

Fig. 6-14   Single, sealed-type ball bearing *(Photo courtesy of New Departure Division, General Motors Corporation)*

Fig. 6-15 Double-row ball bearing *(Photo courtesy of New Departure Division, General Motors Corporation)*

end shield enables maintenance personnel to check the level of the oil in the sleeve bearing.

Figures 6-12 to 6-15 illustrate ball bearing units. In some motors, ball bearings are used instead of sleeve bearings. Grease rather than oil is used to lubricate ball bearings. This type of bearing usually is two-thirds full of grease at the time the motor is assembled. Special fittings are provided on the end bells so that a grease gun can be used to apply additional lubricant to the ball bearing units at periodic intervals.

When lubricating roller bearings, remove the bottom plug so that the old grease is forced out. The manufacturer's specifications for the motor should be consulted for the lubricant grade recommended, the lubrication procedure, and the bearing loads.

PRINCIPLE OF OPERATION OF A SQUIRREL-CAGE MOTOR

As stated in a previous paragraph on the stator construction, the slots of the stator core contain three separate single-phase windings. When three currents 120 electrical degrees apart pass through these windings, a rotating magnetic field results. This field travels around the inside of the stator core. The speed of the rotating magnetic field depends on the number of stator poles and the frequency of the power source. This speed is called the *synchronous speed* and is determined by the formula:

$$\text{Synchronous speed (r/min)} = \frac{120 \times \text{frequency in hertz}}{\text{Number of poles}}$$

$$S = \frac{120 \times f}{p}$$

S = Synchronous speed
f = Hertz (frequency)
p = Number of poles

Example 1. If a three-phase, squirrel-cage induction motor has six poles on the stator winding and is connected to a three-phase, 60-hertz source, then the synchronous speed of the revolving field is 1,200 r/min.

$$S = \frac{120 \times f}{p} = \frac{120 \times 60}{6} = 1,200 \text{ r/min}$$

As this magnetic field rotates at synchronous speed, it cuts the copper bars of the rotor and induces voltages in the bars of the squirrel-cage winding. These induced voltages set up currents in the rotor bars which in turn create a field in the rotor core. This rotor field reacts with the stator field to cause a twisting effect or torque which turns the rotor. The rotor always turns at a speed slightly less than the synchronous speed of the stator field. This means that the stator field will cut the rotor bars. If the rotor turns at the same speed as the stator field, the stator field will not cut the rotor bars and there will be no torque.

Speed Regulation and Percent Slip

The squirrel-cage induction motor has very good speed regulation characteristics (the ratio of difference in speed from no load to full load). Speed performance is measured in terms of percent slip. The synchronous speed of the rotating field of the stator is used as a reference point. Recall that the synchronous speed depends on the number of stator poles and the operating frequency. Since these two quantities remain constant, the synchronous speed also remains constant. If the speed of the rotor at full load is deducted from the synchronous speed of the stator field, the difference is the number of revolutions per minute that the rotor slips behind the rotating field of the stator.

$$\text{Percent Slip} = \frac{\text{synchronous speed} - \text{rotor speed}}{\text{synchronous speed}} \times 100$$

Example 2. If the three-phase, squirrel-cage induction motor used in Example 1 has a synchronous speed of 1,200 r/min and a full-load speed of 1,140 r/min, find the percent of slip.

Synchronous speed (Example 1) = 1,200 r/min
Full-load rotor speed   = 1,140 r/min

$$\text{Percent slip} = \frac{\text{synchronous speed} - \text{rotor speed}}{\text{synchronous speed}} \times 100$$

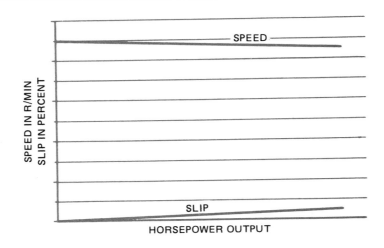

Fig. 6-16 Speed curve and percent slip curve

$$\text{Percent slip} = \frac{1,200 - 1,140}{1,200} \times 100$$

$$\text{Percent slip} = \frac{60}{1,200} \times 100 = .05 \times 100$$

$$\text{Percent slip} = 5 \text{ percent}$$

For a squirrel-cage induction motor, as the value of percent slip decreases, the speed performance of the motor is improved. The average range of percent slip for squirrel-cage induction motors is 2 percent to 6 percent.

Figure 6-16 shows a speed curve and a percent slip curve for a squirrel-cage induction motor operating between no load and full load. The rotor speed at no load slips behind

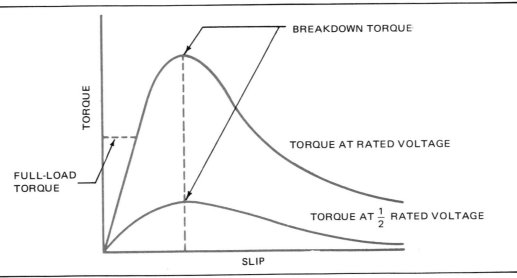

Fig. 6-17 Slip-torque curves for a squirrel-cage motor

the synchronous speed of the rotating stator field just enough to create the torque required to overcome friction and windage losses at no load. As a mechanical load is applied to the motor shaft, the rotor tends to slow down. This means that the stator field (turning at a fixed speed) cuts the rotor bars a greater number of times in a given period. The induced voltages in the rotor bars increase, resulting in more current in the rotor bars and a stronger rotor field. There is a greater magnetic reaction between the stator and rotor fields which causes a stronger twisting effect or torque. This also increases stator current taken from the line. The motor is able to handle the increased mechanical load with very little decrease in the speed of the rotor.

Typical slip-torque curves for a squirrel-cage induction motor are shown in figure 6-17. The torque output of the motor in pound-feet (lb·ft) increases as a straight line with an increase in the value of percent slip as the mechanical load is increased to the point of full load. Beyond full load, the torque curve bends and finally reaches a maximum point called the breakdown torque. If the motor is loaded beyond this point, there will be a corresponding decrease in torque until the point is reached where the motor stalls. However, all induction motors have some slip in order to function.

Starting Current

When a three-phase, squirrel-cage induction motor is connected across the full line voltage, the starting surge of current momentarily reaches as high a value as 400% to 600% of the rated full-load current. At the moment the motor starts, the rotor is at a standstill. At this instant, therefore, the stator field cuts the rotor bars at a faster rate than when the rotor is turning. This means that there will be relatively high induced voltages in the rotor which will cause heavy rotor currents. The resulting input current to the stator windings will be high at the instant of starting. Because of this high starting current, starting protection rated as high as 300 percent of the rated full-load current is provided for squirrel-cage induction motor installations.

Most squirrel-cage induction motors are started at full voltage. If there are any questions concerning the starting of large sizes of motors at full voltage, the electric utility company should be consulted. In the event that the feeders and protective devices of the electric utility are unable to handle the large starting currents, reduced voltage starting circuits must be used with the motor.

Power Factor

The power factor of a squirrel-cage induction motor is poor at no-load and low-load conditions. At no load, the power factor can be as low as 15 percent lagging. However, as load is applied to the motor, the power factor increases. At the rated load, the power factor may be as high as 85 to 90 percent lagging.

The power factor at no load is low because the magnetizing component of input current is a large part of the total input current of the motor. When the load on the motor is increased, the in-phase current supplied to the motor increases, but the magnetizing component of current remains practically the same. This means that the resultant line current is more nearly in phase with the voltage and the power factor is

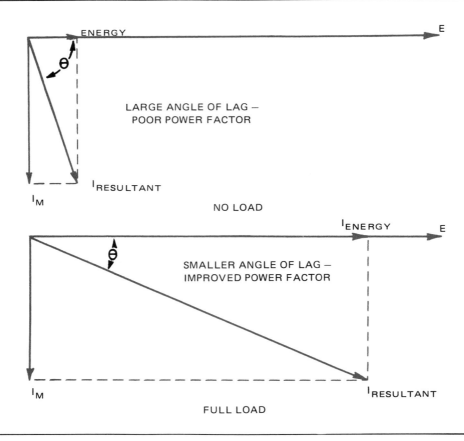

Fig. 6-18 Power factor at no load and full load

improved when the motor is loaded, compared with an unloaded motor which draws its magnetizing current chiefly.

Figure 6-18 shows the increase in power factor from a no-load condition to full load. In the no-load diagram, the in-phase current (I_{energy}) is small when compared to the magnetizing current I_M; thus, the power factor is poor at no load. In the full-load diagram, the in-phase current has increased while the magnetizing current remains the same. As a result, the angle of lag of the line current decreases and the power factor increases.

Reversing Rotation

The direction of rotation of a three-phase induction motor can be reversed readily. The motor will rotate in the opposite direction if any two of the three line leads are reversed (figure 6-19). The leads are reversed at the motor.

Speed Control

A squirrel-cage induction motor has almost no speed control. Recall that the speed of the motor depends on the frequency of the three-phase source and the number of poles of the stator winding.

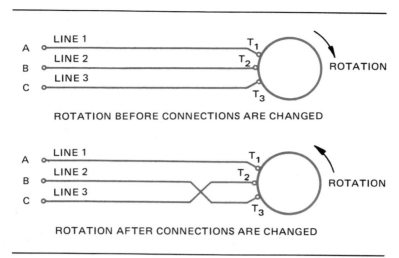

Fig. 6-19 Reversing rotation of an induction motor

The frequency of the supply line is usually 60 hertz, and is maintained at this value by the local power utility company. Since the number of poles in the motor is also a fixed value, the synchronous speed of the motor remains constant. As a result, it is not possible to obtain a range of speed. The three-phase, squirrel-cage induction motor, therefore, is used for applications where a wide range of speed is not necessary.

INDUCTION MOTORS WITH DUAL-VOLTAGE CONNECTIONS

Many three-phase, squirrel-cage induction motors are designed to operate at two different voltage ratings. For example, a typical dual-voltage rating for a three-phase motor is 230/460 volts.

Figure 6-20 shows a typical wye-connected stator winding which may be used for either 230 volts, three phase or 460 volts, three phase. Each of the three single-phase windings consists of two coil windings. There are nine leads brought out externally from this type of stator winding. These leads, identified as leads 1 to 9, end in the terminal box of the motor. To mark the terminals, start at the upper left-hand terminal T_1 and proceed in a clockwise direction in a spiral toward the center, marking each lead as indicated in the figure.

Figure 6-21 shows the connections required to operate a motor from a 460-volt, three-phase source. The two coils of each single-phase winding are connected in series. Figure 6-22 shows the connections to permit operation from a 230-volt, three-phase source.

MOTOR NAMEPLATES

Motor nameplates provide information vital to the proper selection and installation of the motor. Most useful data given on the nameplate refers to the electrical characteristics of the motor. Given this information and using the National Electrical Code, the electrician can determine the conduit, wire, and starting and running protection sizes. (The NEC gives minimum requirements.)

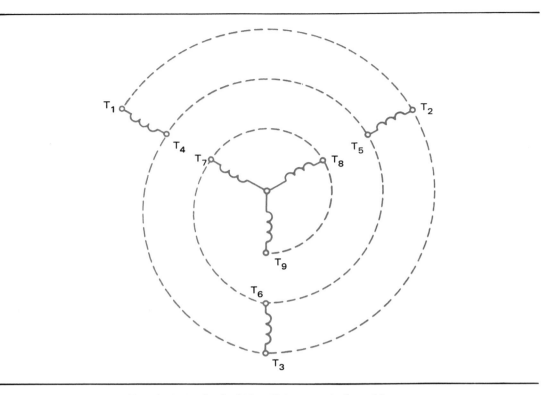

Fig. 6-20 Method of identifying terminal markings

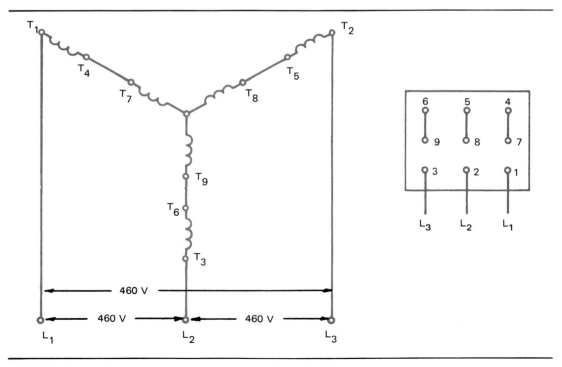

Fig. 6-21 460-volt wye connection. Coils are connected in series.

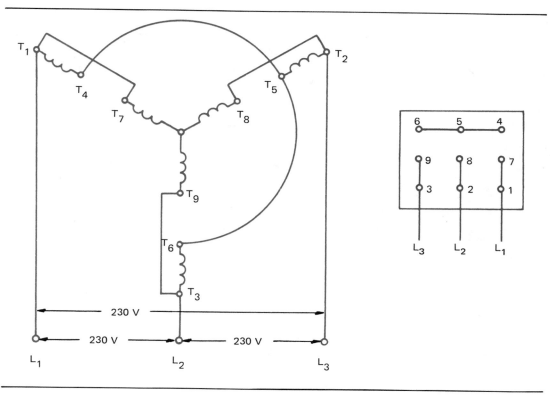

Fig. 6-22 230-volt wye connection. Coils are connected in parallel.

The design and performance data given on the nameplate is useful to maintenance personnel. The information is vital for the fast and proper replacement of the motor, if necessary. For a better understanding of the motor, typical information found on motor nameplates is described as follows (figure 6-23).

- The manufacturer's name

- *Type* identifies the type of the enclosure. This is the manufacturer's coded identification system.

- *Serial number* is the specific motor identification. This is the individual number assigned to the motor, similar to a social security number for a person. It is kept on file by the manufacturer.

- The *model number* is an additional manufacturer's identification, commonly used for ordering purposes.

- *Frame* size identifies the measurements of the motor.

- *Service factor* — a service factor of 1.0 means the motor should not be expected to deliver more than its rated horsepower. The motor will operate safely without injury to the insulation system if it is run at the rated horsepower times the service factor, maximum. Common service factors are 1.0 to 1.15. It is recommended that the motor *not* be run continuously in the service factor range. This may shorten the life expectancy of the insulation system.

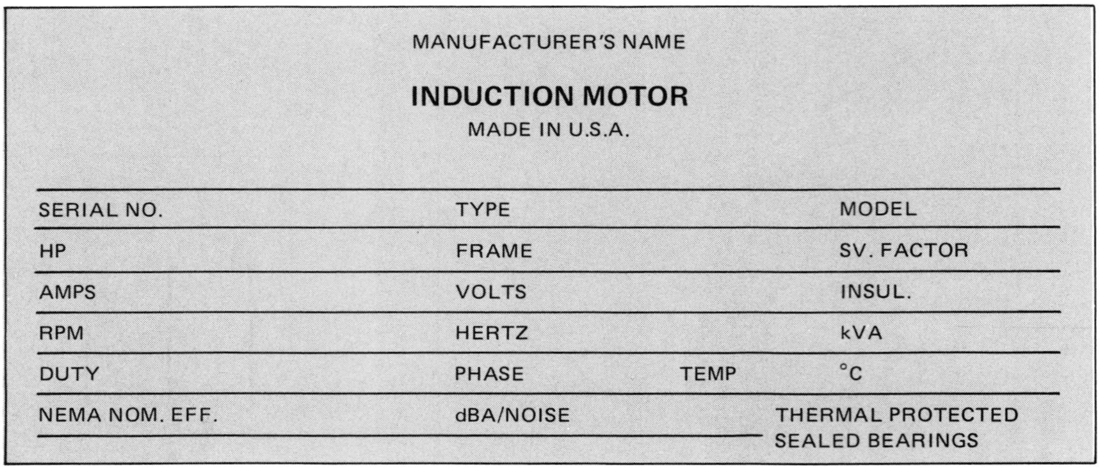

MANUFACTURER'S NAME		
INDUCTION MOTOR		
MADE IN U.S.A.		
SERIAL NO.	TYPE	MODEL
HP	FRAME	SV. FACTOR
AMPS	VOLTS	INSUL.
RPM	HERTZ	kVA
DUTY	PHASE TEMP	°C
NEMA NOM. EFF.	dBA/NOISE	THERMAL PROTECTED
		SEALED BEARINGS

Fig. 6-23 Typical motor nameplate

- *Amperes* means the current drawn from the line when the motor is operating at rated voltage and frequency at the fully rated nameplate horsepower.

- *Volts* should be the value measured at the motor terminals and should be the value for which the motor is designed.

- The *class of insulation* refers to the insulating material used in winding the motor stator. For example, in a Class B system, the maximum operating temperature is 130°C; for Class F, it is 155°C; and for Class H, it is 180°C.

- *RPM* (or *r/min*) means the speed in revolutions per minute when all other nameplate conditions are met.

- *Hertz* is the frequency of the power system for which the motor is designed. Performance will be altered if it is operated at other frequencies.

- *Duty* is the cycle of operation that the motor can safely operate. "Continuous" means that the motor can operate fully loaded 24 hours a day. If "intermediate" is shown, a time interval will also appear. This means the motor can operate at full load for the specified period. The motor should then be stopped and allowed to cool before starting again.

- *Ambient temperature* specifies the maximum surrounding air temperature at which the motor can operate to deliver the rated horsepower.

- *Phase* entry indicates the number of voltage phases at which the motor is designed to operate.

- *kVA* is a code letter which indicates the starting inrush current. This is used to determine starting equipment and protection for the motor. A code table is found in the National Electrical Code.

- *Efficiency* is expressed in percent. This value is found on standard motors as well as "premium efficiency" motors.

- *Noise* — some motors are designed for low noise emission. The noise level given on the nameplate is measured in "dBA" units of sound.
- *Manufacturer's notes* — list specific features of the motors, such as "thermal protected" and/or "sealed bearings."

ALTITUDE

Manufacturers' guarantees for standard motor ratings are usually based on operation at any altitude up to 3,300 feet. Motors suitable for operation at an altitude higher than 3,300 feet above sea level are of special design and/or have a different insulation class. For example, standard motors having a service factor of 1.15 may be operated up to an altitude of 9,900 feet by utilizing the service factor. At an altitude of 9,900 feet, the service factor would be 1.00. It may be necessary to de-rate the motor or to use a larger frame size.

ACHIEVEMENT REVIEW

A. Answer the following statements and questions.

1. List the essential parts of a squirrel-cage induction motor. _____

_____

_____

2. State two advantages of using a squirrel-cage induction motor. _____

_____

_____

3. State two disadvantages of a squirrel-cage induction motor. _____

_____

_____

4. List the two factors which determine the synchronous speed of an induction motor.

_____

_____

_____

5. Explain how to reverse the direction of rotation of a three-phase, squirrel-cage induction motor._____

_____

_____

6. A four-pole, 60-hertz, three-phase, squirrel-cage induction motor has a full-load speed of 1,725 r/min. Determine the synchronous speed of this motor.

_____

_____

_____

7. What is the percent slip of the motor given in question 6? _____

8. Why is the term squirrel-cage applied to this type of three-phase induction motor?

_____

_____

_____

B. Select the correct answer for each of the following statements and place the corresponding letter in the space provided.

9. Who or what determines if large induction motors may be started
   at full voltage across the line?   _____
   a. maximum motor size
   b. rated voltage
   c. the power company
   d. department of building and safety

10. The power factor of a three-phase, squirrel-cage induction motor
    operating unloaded is   _____
    a. the same as with full load.
    b. very poor.
    c. very good.
    d. average.

11. The power factor of a three-phase, squirrel-cage induction motor
    operating with full load   _____
    a. improves from no load.
    b. decreases from no load.
    c. remains the same as at no load.
    d. becomes 100 percent.

12. The squirrel-cage induction motor is popular because of its characteristics of   _____
    a. high percent slip.
    b. low percent slip.
    c. simple, rugged construction.
    d. good speed regulation.

13. The speed of a squirrel-cage induction motor depends on   _____
    a. voltage applied.
    b. frequency and number of poles.
    c. field strength.
    d. current strength.

14. Speed is calculated using the formula   _____

    a. $p = \dfrac{120 \times f}{r/min}$ c. $r/min = \dfrac{p \times f}{120}$

    b. $r/min = \dfrac{120 \times p}{f}$ d. $r/min = \dfrac{120 \times f}{p}$

C. Draw the connection arrangements.

15. Show the connection arrangement for the nine terminal leads of a wye-connected three-phase motor rated at 230/460 volts for operation at 460 volts, three phase.

16. Show the connection arrangement for the nine terminal leads of a wye-connected three-phase motor rated at 230/460 volts for operation at 230 volts, three phase.

UNIT 7
ELECTROMECHANICAL AND SOLID-STATE RELAYS

OBJECTIVES

After studying this unit, the student will be able to

- tell how relays operate.
- list the principal uses of relays.
- describe different relay control and load conditions.
- tell how SCRs operate.
- identify relay component symbols.
- connect different relays in a circuit.

ELECTROMECHANICAL RELAYS

Electromechanical relays, contactors, and motor starters operate basically by the same principles. These electrically-operated switches respond to the electromagnetic attraction of an energized coil of wire in an iron core. The devices differ in the amount of current that each must switch. The relay — which can be compared to an amplifier — is used to switch small amounts of currents (usually 0–15 amperes) in many control circuits (figure 7-1). Uses of relays include switching on and off larger coils of motor starters, contactors, solenoids, heating elements, and small motors. Other uses are in alarm systems and pilot light control. Relays have many industrial and commercial applications, both ac and dc.

A small current flow and/or low voltage applied to a relay coil can result in a much larger current or voltage being switched. One input signal (voltage) may control several output (switched) circuits (figure 7-2).

Fig. 7-1 30-ampere, SPDT power relay for dc or ac operation *(Photo courtesy of Magnecraft Electric Company)*

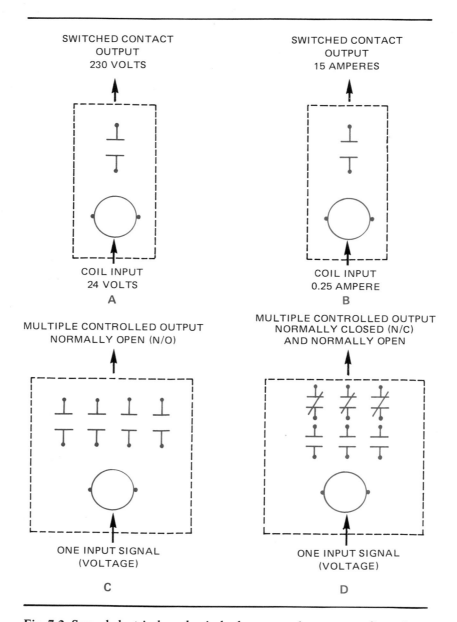

Fig. 7-2 Several electrical-mechanical relay uses and contact configurations

The coil voltages of the relay are separate or different from those at the switched contacts; this is called *separate control*. However, the coil voltage may be the same system voltage as the switched voltage.

Relays are available in many shapes and sizes. Some are sealed in dustproof, transparent plastic enclosures (figure 7-3A). The general construction of a typical relay is shown in figure 7-3B.

Note in figure 7-2 that relay contacts may be normally closed or normally open. The action of the contacts is to switch something "on" or "off" depending upon the configuration.

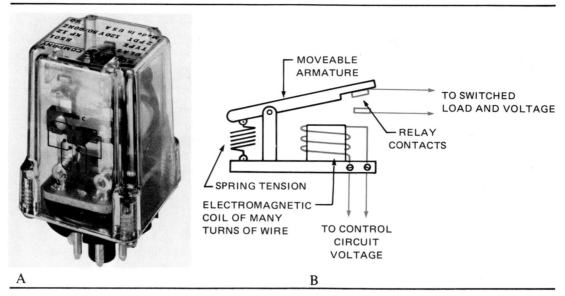

Fig. 7-3 A) Dustproof, transparent enclosure relay B) Construction of a typical electromagnetic control relay *(Photo courtesy of Square D Company)*

SOLID-STATE RELAY

A solid-state relay can be used to control most of the same circuits that the electromechanical relay controls. By comparison, the solid-state relay has no coil or contacts. The semiconductor industry has developed solid-state components with unusual applications to the industrial control processes. These components are very compact, versatile, and becoming more reliable as a result of improvements in materials.

The silicon-controlled rectifier (SCR) is probably the most popular solid-state device for controlling large and small electrical power loads. Basically, the SCR is a rectifier which either conducts or does not conduct an electric current. When it is not conducting, the SCR offers almost a complete blockage to the current. It passes only a few milliamperes to the load. For this reason, some manufacturers place contacts from electrically operated contactors in the total circuit to disconnect the load completely.

The SCR will not conduct when the voltage across it is in the reverse direction. It will conduct only in the forward direction when the proper signal (voltage) is applied to the gate terminal (figure 7-4). Once it is conducting, the SCR cannot be turned off immediately. It is necessary only to provide a small signal to start the SCR conducting a current. It will continue to conduct even without a signal from that point on as long as the current is in the forward direction. The only way to stop the SCR from conducting is to reduce the current flow below the holding current level, or disconnect it from the system. On alternating current, of course, this happens every half-cycle, so this characteristic is no problem (figure 7-5). For direct-current applications, the voltage is reduced to zero by interrupting the circuit, generally with a contact on an electromagnetic relay.

Fig. 7-4 Schematic symbol for an SCR which
is the heart of the solid-state relay

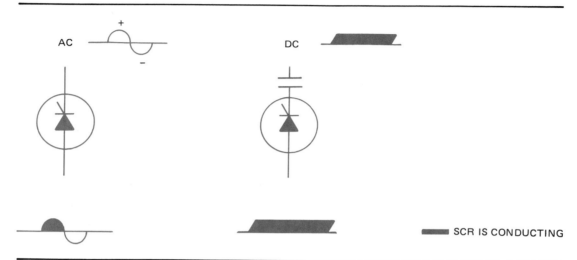

Fig. 7-5 An SCR will conduct current in the forward direction until the voltage is reduced to zero.

Fig. 7-6 Solid-state relay
*(Photo courtesy of
Magnecraft Electric Company)*

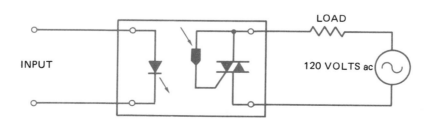

Fig. 7-7 Solid-state relay used to control an ac load. (From Herman, *Electronics for Industrial Electricians*, copyright 1985 by Delmar Publishers Inc.)

Figure 7-6 shows a typical solid-state relay. Note the input and switched output terminal connections. Figure 7-7 shows these connections completed. Terminal wiring is very simple and consists of two input control wires and two output load wires. The connecting terminals are clearly identified on solid-state relays (as they are on electromechanical relays). The relay in figure 7-7 has a light-emitting diode (LED) connected to the input or control voltage. When the input voltage turns the LED on, a photodetector connected to the gate of the triac turns the triac on and connects the load to the line. This optical coupling is commonly used with solid-state relays. These relays are referred to as being *optoisolated*. This means that the load side of the relay is optically isolated from the control side of the relay. The control medium is a light beam. No voltage spikes or electrical noise produced on the load side of the relay are therefore transmitted to the control side.

ACHIEVEMENT REVIEW

1. A mouse trap may be compared with the gate trigger action of an SCR. Indicate whether this statement is true or false and explain your reasoning. _____

   _____

   _____

   _____

   _____

2. What is the major difference between an electromechanical relay and a solid-state relay? _____

   _____

   _____

3. Describe different relay control and load conditions. _____

   _____

   _____

   _____

   _____

4. How are relays used in industrial controls? _____

   _____

   _____

   _____

   _____

5. Basically, what is an SCR? _____

   _____

   _____

   _____

STARTING THREE-PHASE, SQUIRREL-CAGE INDUCTION MOTORS

OBJECTIVES

After studying this unit, the student will be able to

- state the purpose of an across-the line magnetic starting switch.
- describe the basic construction and operation of an across-the-line starter.
- state the ratings for the maximum sizes of fuses required to provide starting protection for motors in the various code marking groups.
- describe what is meant by running overload protection.
- draw a diagram of the connections for an across-the-line magnetic starter with reversing capability.

Alternating-current motors do not require the elaborate starting equipment that must be used with direct-current motors. Most three-phase, squirrel-cage induction motors with ratings up to 10 horsepower are connected directly across the full line voltage. In some cases, motors with ratings greater than 10 horsepower also can be connected directly across the full line voltage. Across-the-line starting usually is accomplished using a magnetic starting switch controlled from a pushbutton station.

The electrician regularly is called upon to install and maintain magnetic motor starters. As a result, the electrician must be very familiar with the connections, operation, and troubleshooting of these starters. The National Electrical Code (NEC) provides information on starting and running overload protection for squirrel-cage induction motors. A comprehensive study of motor controls is given in the text ELECTRIC MOTOR CONTROL.

ACROSS-THE-LINE MAGNETIC STARTER

In the simplest starting arrangement, the three-phase, squirrel-cage motor is connected across full line voltage for operation in one direction of rotation. The magnetic switch used for starting has three heavy contacts, one auxiliary contact, three motor overload relays, and an operating coil. The magnetic switch is called a *motor starter* if it has overload protection. Most motor starters already in service have two overload relays. Three overload relays are now required by the National Electrical Code in new installations.

The wiring diagram for a typical across-the-line magnetic starter is shown in figure 8-1A. The three heavy contacts are in the three line leads feeding the motor. The auxiliary contact acts as a sealing circuit around the normally open start pushbutton

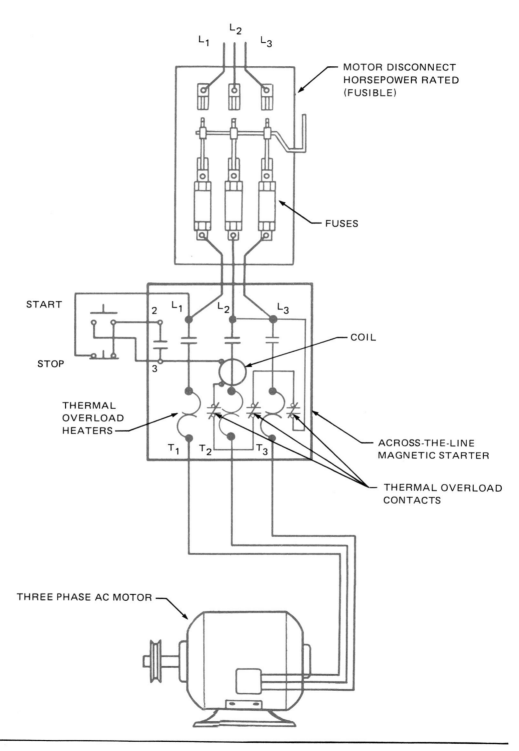

Fig. 8-1A A wiring diagram for an across-the-line magnetic starter

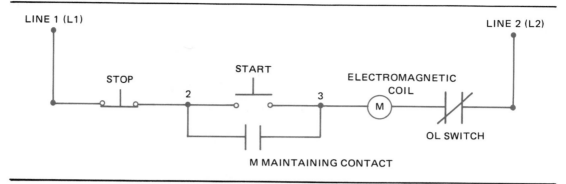

Fig. 8-1B **Elementary diagram of the control circuit for the starter**

Fig. 8-2A   **Start-stop general purpose pushbutton control station** *(Photo courtesy of Square D Company)*

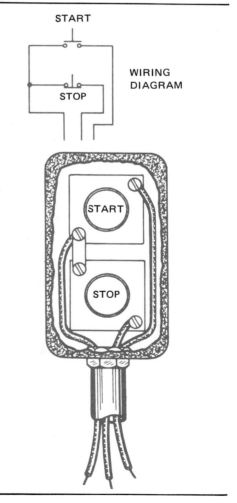

Fig. 8-2B **A pushbutton station and wiring diagram**

Fig. 8-3 Fusible motor disconnect safety switch

when the motor is operating. As a result, the relay remains energized after the start button is released. The four contacts of the across-the-line magnetic starter are operated by the magnetic starter coil controlled from a pushbutton station, as shown in figure 8-1B.

Figure 8-2A shows a typical pushbutton station. Two pushbuttons are housed in a pressed steel box. The start pushbutton is normally open and the stop pushbutton is normally closed, as shown in the diagram (figure 8-2B).

STARTING PROTECTION (BRANCH-CIRCUIT PROTECTION)

In figure 8-1, a motor-rated disconnect switch is installed ahead of the magnetic starter. The safety switch is a three-pole, single-throw enclosed switch. It has a quick-break spring action and is operated externally (figure 8-3). The motor circuit switch contains three cartridge fuses which serve as the starting protection for the motor. These fuses must have sufficient capacity to handle the starting surge of current to the motor. The fuses protect the installation from possible damage resulting from defective wiring

Fig. 8-4 Combination starter with fusible disconnect switch *(Photo courtesy of Square D Company)*

or faults in the motor windings. This combination is available in one enclosure (figure 8-4). (See NEC *Article 430*.)

Briefly, the National Electrical Code gives the following information on starting protection for squirrel-cage induction motors.

1. The maximum size fuses permitted to protect motors with the nameplate code markings from F to V inclusive (with full-voltage starting) are rated at 300 percent of the full-load current of the motor for nontime-delay fuses, and 175 percent for time-delay fuses.
2. The maximum size fuses permitted to protect motors with the nameplate code markings from B to E inclusive are rated at 250 percent of the full-load current of the motor for nontime-delay fuses, and 175 percent for time-delay fuses.
3. The maximum size fuses permitted to protect motors with the nameplate code marking A are rated at 150 percent of the full-load current of the motor for both nontime-delay and time-delay fuses.

The National Electrical Code also covers motors that do not have code letter markings. (Refer to NEC *Table 430-152*.) Note that the maximum size fuses permitted for squirrel-cage motors without code letters are rated at 250 percent of the full-load current of the motor for nontime-delay fuses, and 175 percent for time-delay fuses.

NOTE: If the required fuse size as determined by applying the given percentages does not correspond with the standard sizes of fuses available, and if the specified overcurrent protection is not sufficient to handle the starting current of the motor, then the next higher standard fuse size may be used. In no case can the fuse size exceed 300 percent of the full-load current of the motor for nontime-delay fuses and 175 percent of the full-load current for time-delay fuses. (See the National Electrical Code.)

The marking system for squirrel-cage induction motors was developed by the National Electrical Manufacturers Association (NEMA). Note that the starting surge of current for motors with different code letter identifications varies from 150 percent to 300 percent of the rated full-load current, NEC *Table 430-152*. The differences in the starting current surges are due to differences in the design and construction of the rotor (figure 8-5). An ac magnetic motor starter is shown in figure 8-6A, and an ac reversing magnetic motor starter is shown in figure 8-6B.

Example 1. A three-phase, squirrel-cage induction motor with a nameplate marking of code letter F is rated at 5 hp, 230 volts. This motor has a full-load current per terminal of 15.2 amperes. According to the National Electrical Code, the starting protection shall not exceed 300 percent of the rated current for squirrel-cage motors with code letter markings from F to V inclusive for nontime-delay fuses. Thus, the starting protection is $15.2 \times 3 = 45.6$ amperes.

Since a 45.6-ampere fuse cannot be obtained (see NEC *Section 240-6*), the next largest size of fuse (50 amperes) can be used. For motor branch-circuit protection, the motor current listed in the appropriate table of the National Electrical Code should be used. The full-load current, as stated on the motor nameplate, is not used for this purpose.

INDUCTION MOTOR WITH CODE LETTER A

THIS TYPE OF MOTOR HAS A HIGH-RESISTANCE ROTOR WITH SMALL ROTOR BARS. THIS MOTOR HAS A HIGH STARTING TORQUE AND LOW STARTING CURRENT.

APPLICATIONS:

METAL SHEARS, PUNCH PRESSES, AND METAL DRAWING MACHINERY

INDUCTION MOTOR WITH CODE LETTERS B-E

THIS TYPE OF MOTOR HAS A HIGH-REACTANCE AND LOW-RESISTANCE ROTOR. THIS MOTOR HAS A RELATIVELY LOW STARTING CURRENT AND ONLY FAIR STARTING TORQUE.

APPLICATIONS:

MOTOR-GENERATOR SETS, FANS, BLOWERS, CENTRIFUGAL PUMPS, OR ANY APPLICATION WHERE A HIGH STARTING TORQUE IS NOT REQUIRED.

INDUCTION MOTOR WITH CODE LETTERS F-V

THIS TYPE OF MOTOR HAS A RELATIVELY LOW-RESISTANCE AND LOW-INDUCTIVE REACTANCE ROTOR. THIS MOTOR HAS A HIGH STARTING CURRENT AND ONLY FAIR STARTING TORQUE.

APPLICATIONS:

MOTOR-GENERATOR SETS, FANS, BLOWERS, CENTRIFUGAL PUMPS, OR ANY APPLICATION WHERE A HIGH STARTING TORQUE IS NOT REQUIRED.

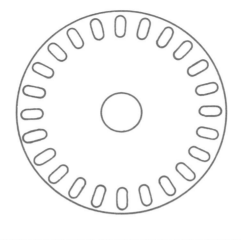

Fig. 8-5 Various types of rotor laminations

A   B

Fig. 8-6 A) Ac magnetic starter such as is used in the combination starter shown in figure 8-4 *(Photo courtesy of Furnas Electric Co.)* B) Ac reversing magnetic motor starter. The elementary diagram of the starter is shown in figure 8-8 *(Photo courtesy of Square D Company)*

RUNNING OVERLOAD PROTECTION

The running overload protection consists of three thermal overload units inserted in series with the three-phase leads feeding to the motor. These overload units generally are located in the magnetic starter. The National Electrical Code requires the use of three thermal overload units as running overload protection. Although new installations require three overload relays, the electrician will work on many older installations which have only two overload relays which were installed before the three overload relay requirement became effective. The overload relay unit may be either an individual unit or a common block containing the three heaters and only one trip switch contact unit reacting from any one of the heaters.

These overload heater units are made of a special alloy. Motor current through these units causes heat to be generated. In one type, a small bimetallic strip is located next to each of the two heater units. When an overload on a motor continues for a period of approximately one to two minutes, the excessive heat developed by the heater units causes the bimetallic strips to expand. As each bimetallic strip expands, it causes the normally closed contacts in the control circuit to open. The main relay coil is deenergized and disconnects the motor by opening the main and auxiliary contacts. Melting alloy heaters (solder pots) also are commonly used. The heat generated by the overload melts the sealed solder pot to a release ratchet which trips the control circuit contacts.

Before the motor can be restarted at the pushbutton station, the overload contacts in the control circuit *must be allowed to cool* before being reclosed (reset). When the reset button in the magnetic starter is pressed, the overload contacts in the control

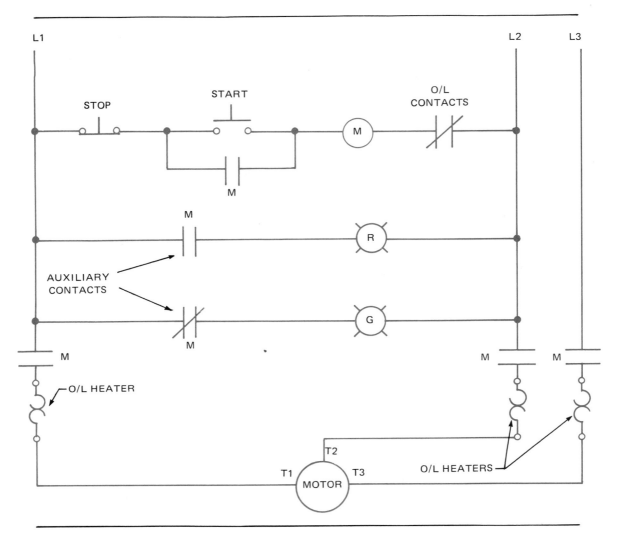

Fig. 8-7 Electrical interlocks (auxiliary contacts) switch pilot lights in this circuit

circuit are reset to their normally closed position. The motor then can be controlled from the pushbutton station.

The National Electrical Code requires that the running overload protection in each phase be rated at not more than 125 percent of the full-load current rating for motors which are indicated to have a temperature rise of not more than 40 degrees Celsius (see NEC *Article 430, Part C*).

Example 2. Using the motor full-load current rating from the nameplate data, determine the running overcurrent protection for a three-phase, 5-hp, 230-volt squirrel-cage induction motor with a rated full-load current of 14.5 amperes and a temperature rise of 40 degrees Celsius. The running overcurrent protection is 14.5 × 1.25 = 18.1 amperes.

For this motor, heater overload units rated at 18.5 amperes are required for the magnetic starter. Where the overload relay so selected is not sufficient to start this motor, the next higher size overload relay is permitted, but not to exceed 140 percent of the motor full-load current rating. Actual motor nameplate currents are used to establish the overload protection.

AUXILIARY CONTACTS

In addition to the standard contacts, a starter can be provided with externally attached auxiliary contacts, sometimes called *electrical interlocks* (figure 8-7). These auxiliary contacts can be used in addition to the holding circuit contacts, and the main or power contacts which carry the motor current. Auxiliary contacts are rated to carry only control circuit currents of 0–15 amperes, not motor currents. Versions are available with either normally open or normally closed contacts. Among a wide variety of applications, auxiliary contacts are used to

- control other magnetic devices where sequence operation is desired.
- electrically prevent another controller from becoming energized at the same time (such as reverse starting), called interlocking.
- make and break circuits of indicating or alarm devices, such as pilot lights, bells, or other signals.

Auxiliary contacts are packaged in kit form, and can be added easily in the field.

ACROSS-THE-LINE MOTOR STARTER WITH REVERSING CAPABILITY

The direction of rotation of a squirrel-cage induction motor must be reversed for some industrial applications. To reverse the direction of rotation, interchange any two of the three line leads.

Figure 8-8 is an elementary wiring diagram of a motor starter having a reversing capability. When the three main reverse contacts are closed, the phase sequence at the motor terminals is different from that when the three main forward contacts are closed. Two of the line leads feeding to the motor are interchanged when the three reverse contacts close.

The control circuit has a pushbutton station with Forward, Reverse, and Stop pushbuttons. The control circuit requires a mechanical and an electrical interlocking system provided by the push buttons. Electrical interlocking means that if one of the devices in the control circuit is energized, the circuit to a second device is open and cannot be closed until the first device is disconnected. Mechanical interlocks, shown by the broken lines in figure 8-8, are used between the forward and reverse coils and pushbuttons.

Note in figure 8-8 that when the forward pushbutton is pressed, it breaks contact with terminals 4 and 5 opening the reverse coil circuit, and makes contact between terminals 4 and 7. As a result, coil F is energized and the forward contacts close. The motor now rotates in the forward direction. If the reverse pushbutton is pressed, it will break contact between terminals 7 and 8 and open the circuit to coil F. This causes all forward contacts to open. As the reverse pushbutton is depressed more, it closes the contact between terminals 5 and 6 and energizes coil R. All reverse contacts

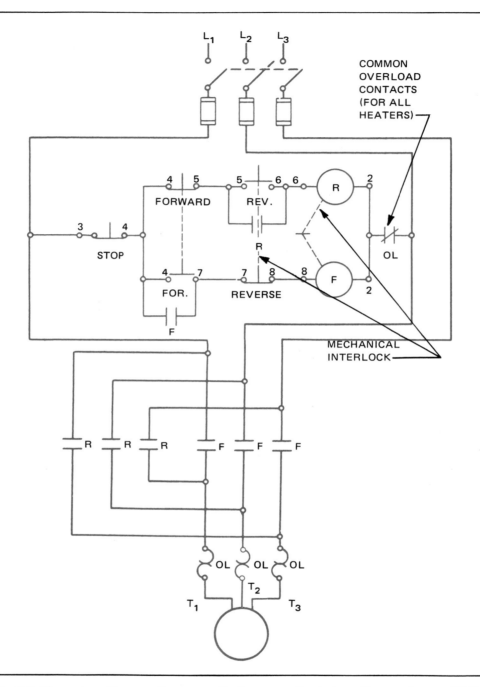

Fig. 8-8 Elementary diagram of an across-the-line magnetic starter with reversing capabilities

are now closed and the motor rotates in the reverse direction. If the stop button is pressed, the contact between terminals 3 and 4 is opened, the control circuit is interrupted, and the motor is disconnected from the three-phase source.

Figure 8-9 is a panel wiring diagram of the motor starter shown in figure 8-8. The National Electrical Code requirements for starting and running overload protection

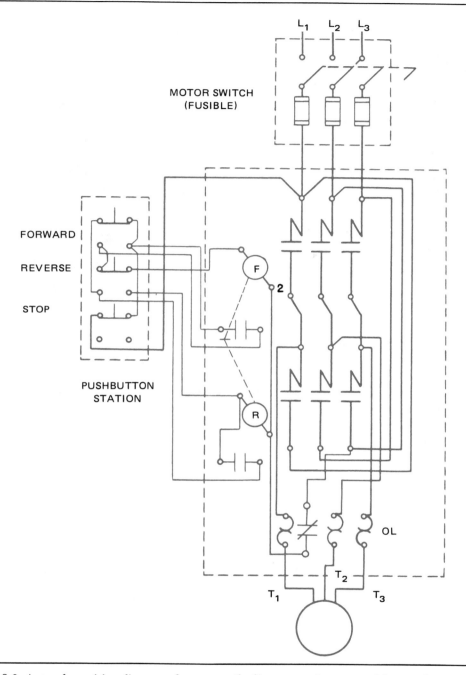

Fig. 8-9 A panel or wiring diagram of an across-the-line magnetic starter with reversing capability

which apply to the across-the-line motor starter also apply to this type of motor starter.

DRUM REVERSING SWITCH

A drum reversing switch (figure 8-10) may be used to reverse the direction of rotation of squirrel-cage induction motors.

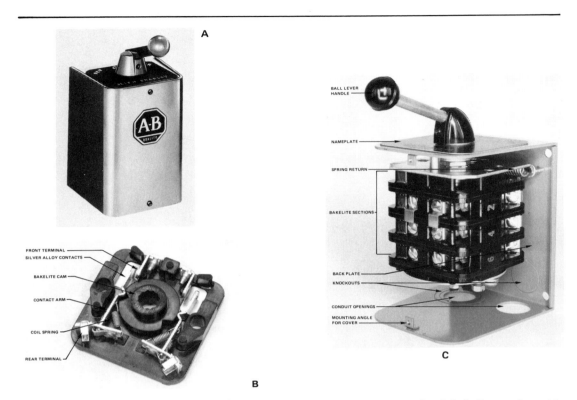

Fig. 8-10 A) Reversing drum switch B) A bakelite section of a drum switch C) Bakelite section with cover removed *(Photos courtesy of Allen-Bradley Company)*

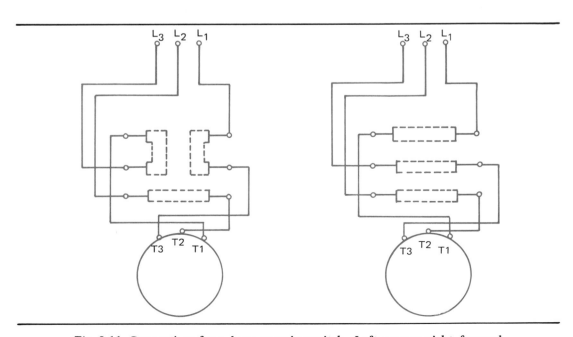

Fig. 8-11 Connections for a drum reversing switch. Left, reverse; right, forward

The motor is started in the forward direction by moving the handle of the drum reversing switch from the off position to the forward (F) position. The connections for this drum controller in both the forward and reverse positions are shown in figure 8-11. In the forward position, the switch connects line 1 to motor terminal 1, line 2 to motor terminal 2, and line 3 to motor terminal 3.

To reverse the direction of rotation, the drum switch handle is moved to the reverse (R) position. In the reverse position, line 1 is still connected to motor terminal 1. However, line 2 is now connected to motor terminal 3, and line 3 is connected to motor terminal 2. When the handle of the drum switch is moved to the off position, all three line leads are disconnected from the motor.

ACHIEVEMENT REVIEW

1. What is the purpose of starting protection for a three-phase motor? _____

   _____

   _____

2. What is the purpose of running overload protection for a three-phase motor?

   _____

   _____

3. What is meant by the code letter markings of squirrel-cage induction motors?

   _____

   _____

4. List some of the industrial applications for squirrel-cage induction motors with code letter classification A. _____

   _____

5. List some of the industrial applications for squirrel-cage induction motors with code letter classifications B to E. _____

   _____

6. List some of the industrial applications for squirrel-cage induction motors with code letter classifications F to V. _____

   _____

7. A three-phase motor (code letter J) has a full-load current rating of 40 amperes, and a temperature rise of 40°C.
   a. What are the maximum size fuses that can be used for branch-circuit protection?

      _____

      _____

   b. What size heaters would be used for running overcurrent protection?

      _____

      _____

8. What is the maximum starting protection allowed by the National Electrical Code?

_____

_____

MANUAL STARTING COMPENSATOR

OBJECTIVES

After studying this unit, the student will be able to

- describe the operation of a manual starting compensator used to start a large three-phase, squirrel-cage induction motor.
- connect a manual starting compensator to a three-phase, squirrel-cage motor.
- describe a common method of supplying running overload protection to a motor with a starting compensator.
- state how starting overload protection is provided for a motor-starting compensator combination.

MANUAL STARTING COMPENSATOR

Large three-phase, squirrel-cage induction motors normally are started by means of starting compensators (figure 9-1). A manual starting compensator has a Start position and a Run position. With the compensator in the start position, the three-phase

Fig. 9-1 Manual reduced voltage starting compensator with air break contact construction *(Photo courtesy of Allen-Bradley Company)*

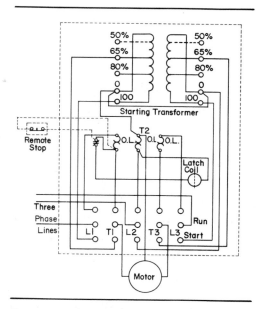

Fig. 9-2 Wiring diagram of the manual starting compensator shown in figure 9-1 *(Courtesy of Allen-Bradley Company)*

motor is connected across 45 percent of the rated line voltage. In this way, the starting surge of current of the motor is reduced. Many automatic compensators have provisions for connecting three-phase motors across 45 percent, 65 percent, 80 percent and 100 percent of the rated line voltage.

Figure 9-2 shows the internal connections of a starting compensator. The compensator has two sets of stationary contacts and one set of movable contacts. The movable contacts are mounted on an insulated cylinder which is attached to the external handle. When the motor is started, the handle is moved quickly to the start position. The three autotransformers shown in the figure are first connected in wye and then are connected to the three-phase line. Connections made to the 50 percent taps on the three autotransformers feed to the motor terminals. As a result, only 50 percent of the line voltage is applied to the motor terminals. The handle is held in the start position by the operator until the motor accelerates to a speed which is close to its rated speed. At this point, the handle is *pulled rapidly* to the run position. The motor is disconnected from the autotransformer for an instant and reconnected directly across the three-phase line. This connection is called "open transition." Note that the autotransformers are disconnected from the line with the handle in the run position. Once the handle is in the run position, the undervoltage coil which is connected across two of the line leads secures the operating handle mechanism in place.

To stop the motor, the stop button is pressed. This action opens the circuit to the undervoltage release coil and the handle mechanism is released. A spring action insures that the movable contacts and the compensator handle move quickly to the off position.

Automatic compensators are available which use timing relays and contactors to start motors in a manner similar to that used by manual starters. For detailed information, refer to the text, ELECTRIC MOTOR CONTROL.

Most manual starting compensators are made with the stationary and movable contacts immersed in an insulating oil (figure 9-3). The insulating oil minimizes any arcing which may occur between the stationary and movable contacts. In this way, the contacts are prevented from wearing rapidly because of severe pitting and burning. The contacts must be inspected periodically by the electrician to determine if they are in good condition. Often, it is necessary to dress the contacts with a fine file or sandpaper to remove rough, pitted surfaces. Silver contacts, however, should not be dressed. The electrician should also check to insure that the movable and stationary contacts close properly. Since the insulating oil becomes carbonized after the starting compensator has been in active use for a period of time, the oil must be replaced at periodic intervals.

Fig. 9-3 Oil immersed manual reduced voltage motor starter. The oil tank is lowered for better visibility. *(Photo courtesy of Allen-Bradley Company)*

RUNNING OVERLOAD PROTECTION

The running overload protection consists of three overload units inserted in series with the line wires feeding to the motor. The National Electrical Code requires that three overload units be used to provide running protection on new installations.

Most units already in service have two overload relays. Figure 9-2 indicates that three overload units are used. If the motor is subjected to a sufficient overload, the current flow through the heater coils of the overload units increases. The heaters then actuate the normally closed contact to open the circuit to the no-voltage release latch coil. When the no-voltage release coil is deenergized, the switch contacts and external operating handle are released to the off position.

Melting alloy solder pot heater units also can be used to provide the running overload protection for starting compensators. Thermal overload heater units are made of a special metal alloy. Heat is generated by the passage of the motor current through these units. A small bimetallic strip is mounted next to each of the three heater units. When an overload continues on a motor for approximately one minute, the excessive heat developed by the heater units causes the bimetallic strips to expand. As each bimetallic strip expands, it opens the normally closed contacts in the control circuit.

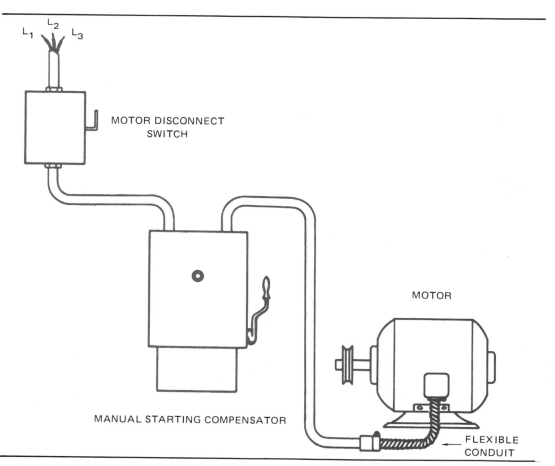

Fig. 9-4 Diagram of a motor installation

This action deenergizes the main relay coil so that the movable switch contacts and the operating handle are released to the off position. After cooling the motor, and before it can be restarted, the overload contacts must be reset to their normally closed position.

Regardless of whether magnetic units or thermal units are used for running overload protection, the National Electrical Code requirements are the same. The Code states that the running overload protection shall be rated at not more than 125 percent of the motor full-load current rating for motors marked to have a temperature rise of not more than 40 degrees Celsius. For other types of motors, consult the National Electrical Code, *Article 430.*

STARTING PROTECTION

Starting overload protection is provided to protect the motor installation from potential damage due to defective wiring or faults in the starting compensator or the motor windings. The starting protection usually consists of a motor disconnect switch containing fuses. The motor disconnect switch is installed ahead of the starting compensator (figure 9-4).

The motor disconnect switch is a three-fuse, three-pole, single-throw enclosed safety switch. It has a quick-break spring action and is operated externally. The three cartridge fuses provide the starting overload protection.

For starting protection requirements, consult the latest edition of the National Electrical Code.

ACHIEVEMENT REVIEW

A.   Give complete answers to questions 1 to 4.

1. What is the main advantage of using a starting compensator with a squirrel-cage induction motor? _____

_____

2. Where will the electrician find most of the maintenance and repair problems on a starting compensator? _____

_____

_____

_____

_____

3. A three-phase, 20-hp squirrel-cage induction motor is rated at 58 amperes per terminal. The temperature rise marked on the nameplate is 40 degrees Celsius. The code letter marking on the nameplate is H. Determine the size of the fuses required as starting protection for this motor when used with a starting compensator.

_____

_____

_____

_____

4. Determine the current setting of thermal overload units in a starting compensator which is to be used with the motor in problem 3. The thermal overload units are to serve as the running overload protection. _____

_____

_____

_____

_____

B. Insert the word or phrase to complete statements 5 through 11.

5. A squirrel-cage induction motor has a code letter marking A. It is used with a starting compensator and must have starting protection not to exceed _____ percent of the full-load current of the motor.

6. A squirrel-cage induction motor with a code letter marking from B to E inclusive is used with a starting compensator. It must have starting protection not to exceed _____ percent of the full-load current of the motor.

7. A squirrel-cage induction motor with a code letter marking from F to V inclusive is used with a starting compensator. It must have starting protection not to exceed _____ percent of the full-load current of the motor.

8. A squirrel-cage induction motor has a nameplate marking for a 40 degree Celsius temperature rise. This motor must have running overload protection not to exceed _____ percent of the full-load current of the motor.

9. A squirrel-cage induction motor marked for a temperature rise of 50 degrees Celsius must have running overload protection not to exceed _____ percent of the full-load current of the motor.

10. On new installations, running overcurrent protection must be provided in _____ line of the three phases.

11. Refer to the diagram of a motor installation in figure 8-3. In the space provided, indicate the proper electrical protection for
    a. the motor disconnect switch.

    _____

    b. the manual starting compensator.

    _____

THREE-PHASE, WOUND-ROTOR INDUCTION MOTOR

OBJECTIVES

After studying this unit, the student will be able to

- list the main components of a wound-rotor, polyphase induction motor.
- describe how the synchronous speed is developed in this type of motor.
- describe how a speed controller connected to the brushes of the motor provides a variable speed range for the motor.
- state how the torque, speed regulation, and operating efficiency of the motor are affected by the speed controller.
- demonstrate how to reverse the direction of rotation of a wound-rotor induction motor.

Many industrial motor applications require three-phase motors with variable speed control. The squirrel-cage induction motor cannot be used for variable speed work since its synchronous speed is essentially constant. Therefore, another type of induction motor was developed for variable speed applications. This motor is called the *wound-rotor induction motor* or *slip-ring ac motor*.

CONSTRUCTION DETAILS

A three-phase, wound-rotor induction motor consists of a stator core with a three-phase winding, a wound rotor with slip rings, brushes and brush holders, and two end shields to house the bearings that support the rotor shaft.

Figures 10-1, 10-2, 10-3, and 10-4 show the basic parts of a three-phase, wound-rotor induction motor.

The Stator

A typical stator contains a three-phase winding held in place in the slots of a laminated steel core, figure 10-2. The winding consists of formed coils arranged and connected so that there are three single-phase windings spaced 120 electrical degrees apart. The separate single-phase windings are connected either in wye or delta. Three line leads are brought out to a terminal box mounted on the frame of the motor. This is the same construction as the squirrel-cage motor stator.

The Rotor

The rotor consists of a cylindrical core composed of steel laminations. Slots cut into the cylindrical core hold the formed coils of wire for the rotor winding.

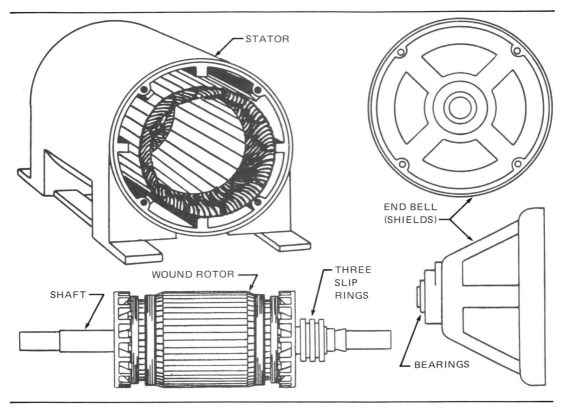

Fig. 10-1 Parts of a wound-rotor motor

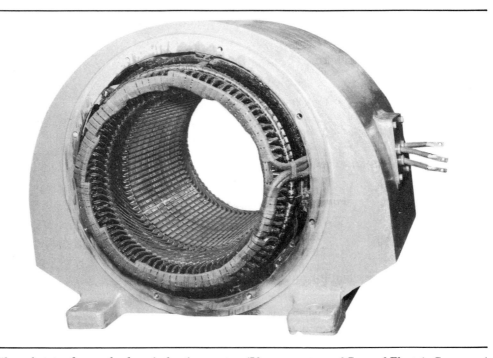

Fig. 10-2 Wound stator for a polyphase induction motor *(Photo courtesy of General Electric Company)*

Fig. 10-3 Wound rotor for a polyphase induction motor *(Photo courtesy of General Electric Company)*

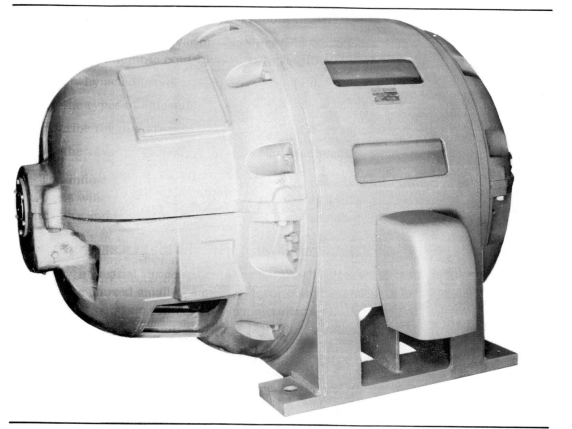

Fig. 10-4 Sleeve bearing, wound-rotor polyphase induction motor *(Photo courtesy of General Electric Company)*

The rotor winding consists of three single-phase windings spaced 120 electrical degrees apart. The single-phase windings are connected either in wye or delta. (The rotor winding must have the same number of poles as the stator winding.) The three leads from the three-phase rotor winding terminate at three slip rings mounted on the rotor shaft. Leads from carbon brushes which ride on these slip rings are connected to an external speed controller to vary the rotor resistance for speed control.

The brushes are held securely to the slip rings of the wound rotor by adjustable springs mounted in the brush holders. The brush holders are fixed in one position. For this type of motor, it is not necessary to shift the brush position as is sometimes required in direct-current generator and motor work.

The Motor Frame

The motor frame is made of cast steel. The stator core is pressed directly into the frame. Two end shields are bolted to the cast steel frame. One of the end shields is larger than the other because it must house the brush holders and brushes which ride on the slip rings of the wound rotor. In addition, it often contains removable inspection covers.

The bearing arrangement is the same as that used in squirrel-cage induction motors. Either sleeve bearings or ball-bearing units are used in the end shields.

PRINCIPLE OF OPERATION

When three currents, 120 electrical degrees apart, pass through the three single-phase windings in the slots of the stator core, a rotating magnetic field is developed. This field travels around the stator. The speed of the rotating field depends on the number of stator poles and the frequency of the power source. This speed is called the synchronous speed. It is determined by applying the formula which was used to find the synchronous speed of the rotating field of squirrel-cage induction motors.

$$\text{Synchronous speed in r/min} = \frac{120 \times \text{frequency in hertz}}{\text{number of poles}}$$

$$S = \frac{120 \times f}{p}$$

As the rotating field travels at synchronous speed, it cuts the three-phase winding of the rotor and induces voltages in this winding. The rotor winding is connected to the three slip rings mounted on the rotor shaft. The brushes riding on the slip rings connect to an external wye-connected group of resistors (speed controller), figure 10-5. The induced voltages in the rotor windings set up currents which follow a closed path from the rotor winding to the wye-connected speed controller. The rotor currents create a magnetic field in the rotor core based on transformer action. This rotor field reacts with the stator field to develop the torque which causes the rotor to turn. The speed controller is sometimes called the *secondary resistance control.*

Starting Theory of Wound-Rotor Induction Motors

To start the motor, all of the resistance of the wye-connected speed controller is inserted in the rotor circuit. The stator circuit is energized from the three-phase line.

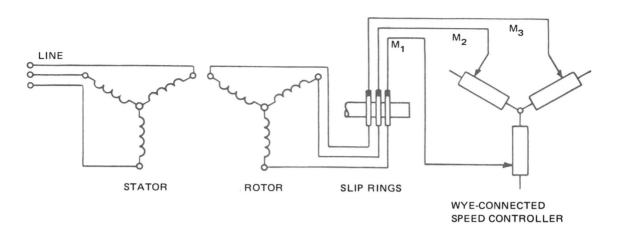

Fig. 10-5 Connections for a wound-rotor induction motor and a speed controller

The voltage induced in the rotor develops currents in the rotor circuit. The rotor currents, however, are limited in value by the resistance of the speed controller. As a result, the stator current also is limited in value. In other words, to minimize the starting surge of current to a wound-rotor induction motor, insert the full resistance of the speed controller in the rotor circuit. As the motor accelerates, steps of resistance in the wye-connected speed controller can be cut out of the rotor circuit until the motor accelerates to its rated speed.

Speed Control

The insertion of resistance in the rotor circuit not only limits the starting surge of current, but also produces a high starting torque and provides a means of adjusting the speed. If the full resistance of the speed controller is cut into the rotor circuit when the motor is running, the rotor current decreases and the motor slows down. As the rotor speed decreases, more voltage is induced in the rotor windings and more rotor current is developed to create the necessary torque at the reduced speed.

If all of the resistance is removed from the rotor circuit, the current and the motor speed will increase. However, the rotor speed always will be less than the synchronous speed of the field developed by the stator windings. Recall that this fact also is true of the squirrel-cage induction motor. The speed of a wound-rotor motor can be controlled manually or automatically with timing relays, contactors, and pushbutton speed selection.

Torque Performance

As a load is applied to the motor, both the percent slip of the rotor and the torque developed in the rotor increase. As shown in the graph in figure 10-6, the relationship between the torque and percent slip is practically a straight line.

Figure 10-6 illustrates that the torque performance of a wound-rotor induction motor is good whenever the full resistance of the speed controller is inserted in the rotor circuit. The large amount of resistance in the rotor circuit causes the rotor current to be almost in phase with the induced voltage of the rotor. As a result, the field set

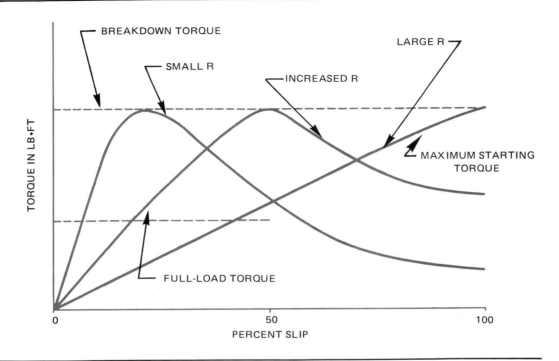

Fig. 10-6 Performance curves of a wound-rotor motor

up by the rotor current is almost in phase with the stator field. If the two fields reach a maximum value at the same instant, there will be a strong magnetic reaction resulting in a high torque output.

However, if all of the speed controller resistance is removed from the rotor circuit and the motor is started, the torque performance is poor. The rotor circuit minus the speed controller resistance consists largely of inductive reactance. This means that the rotor current lags behind the induced voltage of the rotor and, thus, the rotor current lags behind the stator current. As a result, the rotor field set up by the rotor current lags behind the stator field which is set up by the stator current. The resulting magnetic reaction of the two fields is relatively small since they reach their maximum values at different points. In summary, then, the starting torque output of a wound-rotor induction motor is poor when all resistance is removed from the rotor circuit.

Speed Regulation

It was shown in the previous paragraphs that the insertion of resistance at the speed controller improves the starting torque of a wound-rotor motor at low speeds. However, there is an opposite effect at normal speeds. In other words, the speed regulation of the motor is poorer when resistance is added in the rotor circuit at a higher speed. For this reason, the resistance of the speed controller is removed as the motor comes up to its rated speed.

Figure 10-7 shows the speed performance of a wound-rotor induction motor. Note that the speed characteristic curve resulting when all of the resistance is cut out of the

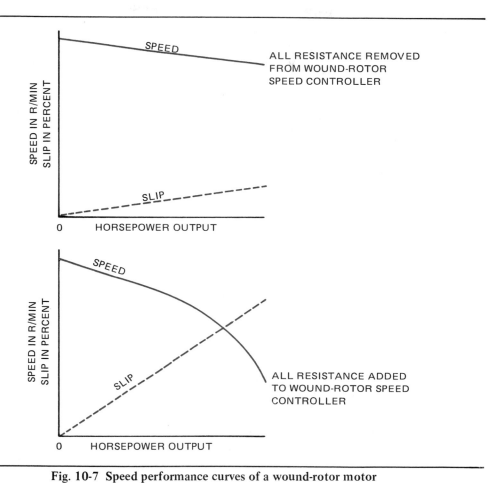

Fig. 10-7 Speed performance curves of a wound-rotor motor

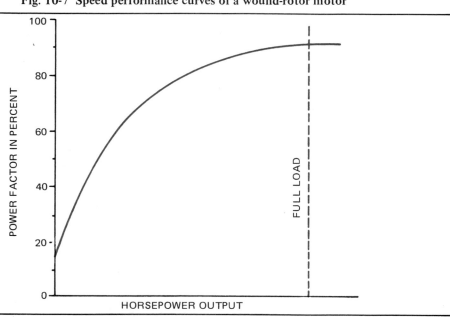

Fig. 10-8 Power factor curve of a wound-rotor induction motor

speed controller indicates relatively good speed regulation. The second speed characteristic curve, resulting when all of the resistance is inserted in the speed controller, has a marked drop in speed as the load increases. This indicates poor speed regulation.

Power Factor

The power factor of a wound-rotor induction motor at no load is as low as 15 percent to 20 percent lag. However, as load is applied to the motor, the power factor improves and increases to 85 percent to 90 percent lag at rated load.

Figure 10-8 is a graph of the power factor performance of a wound-rotor induction motor from a no-load condition to full load. The low lagging power factor at no load is due to the fact that the magnetizing component of load current is such a large part of the total motor current. The magnetizing component of load current magnetizes the iron, causing interaction between the rotor and the stator, by mutual inductance.

As the mechanical load on the motor increases, the in-phase component of current increases to supply the increased power demands. The magnetizing component of the current remains the same, however. Since the total motor current is now more nearly in phase with the line voltage, there is an improvement in the power factor.

Operating Efficiency

Both a wound-rotor induction motor with all of the resistance cut out of the speed controller and a squirrel-cage induction motor show nearly the same efficiency performance. However, when a motor must operate at slow speeds with all of the resistance cut in the rotor circuit, the efficiency of the motor is poor because of the power loss in watts in the resistors of the speed controller.

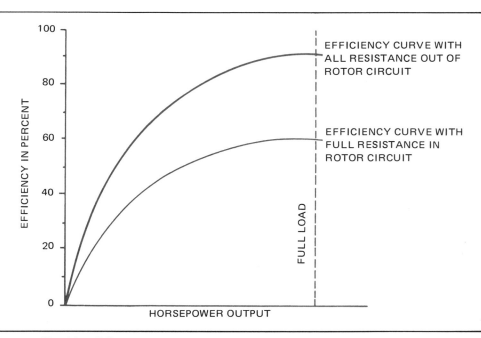

Fig. 10-9 Efficiency curves for a wound-rotor induction motor

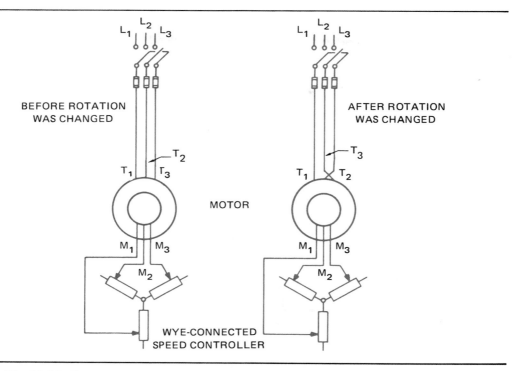

Fig. 10-10 Changes necessary to reverse direction of rotation of a wound-rotor motor

Figure 10-9 illustrates the efficiency performance of a wound-rotor induction motor. The upper curve showing the highest operating efficiency results when the speed controller is in the fast position and there is no resistance inserted in the rotor circuit. The lower curve shows a lower operating efficiency. This occurs when the speed controller is in the slow position and all of the controller resistance is inserted in the rotor circuit.

Reversing Rotation

The direction of rotation of a wound-rotor induction motor is reversed by interchanging the connections of any two of the three line leads, figure 10-10. This procedure is identical to the procedure used to reverse the direction of rotation of a squirrel-cage induction motor.

The electrician should never attempt to reverse the direction of rotation of a wound-rotor induction motor by interchanging any of the leads feeding from the slip rings to the speed controller. Changes in these connections will not reverse the direction of rotation of the motor.

ACHIEVEMENT REVIEW

A.   Give complete answers to the following questions.

1.   List the essential parts of a wound-rotor induction motor. _____

_____

_____

_____

2. List two reasons why a wound-rotor induction motor is started with all of the resistance inserted in the speed controller. _____

    _____

    _____

3. A three-phase, wound-rotor induction motor has six poles and is rated at 60 hertz. The full-load speed of this motor with all of the resistance cut out of the speed controller is 1,120 r/min. What is the synchronous speed of the field set up by the stator windings? _____

    _____

    _____

4. Determine the percent slip at the rated load for the motor in question 3.

    _____

    _____

5. Why is a wound-rotor induction motor used in place of a squirrel-cage induction motor for some industrial applications? _____

    _____

    _____

6. Why is the percent efficiency of a wound-rotor induction motor poor when operating at rated load with all of the resistance inserted in the speed controller?

    _____

    _____

7. What must be done to reverse the direction of rotation of a wound-rotor induction motor? _____

    _____

    _____

8. Why is the power factor of a wound-rotor induction motor poor at no load?

    _____

    _____

    _____

9. List the two factors which affect the synchronous speed of the rotating magnetic field set up by the current in the stator windings. _____

_____

_____

_____

B. Select the correct answer for each of the following statements and place the corresponding letter in the space provided.

10. The speed of a wound-rotor motor is increased by   _____
   a. inserting resistance in the primary circuit.
   b. inserting resistance in the secondary circuit.
   c. decreasing the resistance in the secondary circuit.
   d. decreasing the resistance in the primary circuit.

11. The starting current of a wound-rotor induction motor is limited by   _____
   a. decreasing the resistance in the primary circuit.
   b. decreasing resistance in the secondary circuit.
   c. inserting resistance in the primary circuit.
   d. inserting resistance in the secondary circuit.

12. The direction of rotation of a wound-rotor motor is changed by interchanging any two of the three: _____
   a. L_1, L_2, or L_3.   c. M_1, M_2, or M_3.
   b. T_1, T_2, or T_3.   d. all of these.

13. Wound-rotor motors can be used with   _____
   a. manual speed controllers.
   b. automatic speed controllers.
   c. pushbutton speed selection.
   d. all of these.

14. The full-load efficiency of a wound-rotor motor is best when   _____
   a. all of the resistance is cut out of the secondary circuit.
   b. all of the resistance is cut in the secondary circuit.
   c. it is running slowly.
   d. it is running at medium speed.

15. The main advantage of the wound-rotor polyphase motor is that it   _____
   a. has a low starting torque.   c. will reverse rapidly.
   b. has a wide speed range.   d. has a low speed range.

16. The wound-rotor motor is so-named because the   _____
   a. rotor is wound with wire.
   b. stator is wound with wire.
   c. controller is wound with wire.
   d. all of these.

17. The magnetizing component of load current   _____
    a. is a small part of the total motor current at no load.
    b. magnetizes the iron, causing interaction between the rotor and
        the stator.
    c. is a large part of the total motor current at full load.
    d. is unrelated to the power factor.

MANUAL SPEED CONTROLLERS FOR WOUND-ROTOR INDUCTION MOTORS

OBJECTIVES

After studying this unit, the student will be able to

- state three reasons why speed controllers are used with wound-rotor induction motors.
- list and describe the physical construction of two basic types of manual speed controllers.
- explain the operation of a faceplate controller with a protective starting device.
- explain the operation of a drum controller.
- summarize the National Electrical Code regulations regarding the wire size for the stator circuit, the wire size for the rotor circuit, starting overload protection, and running overload protection.
- draw wiring diagrams for wound rotor motor control.

Many industrial applications require the use of wound-rotor induction motors with speed controllers. This unit covers the details of the operation of manual speed controllers and their connection to wound-rotor motors. Information on National Electrical Code regulations which apply to wound-rotor induction motor installations also is presented.

Reasons for Use of Speed Controllers

Speed controllers are used with wound-rotor induction motors for three basic reasons:

- to limit the starting surge of current to the motor by inserting resistance in the rotor circuit.
- to improve the starting torque of a wound-rotor induction motor by inserting resistance in the rotor circuit.
- to control the speed of a wound-rotor induction motor by varying the resistance in the rotor circuit.

FACEPLATE CONTROLLER

The simplest form of manual speed controller is the *faceplate controller* (figure 11-1. In this type of controller, three sets of contact buttons are mounted on a panel. Each set of contact buttons is connected to a separate tapped resistor housed in the

Fig. 11-1 Speed regulating rheostat with auxiliary control switch for interlocking with or controlling operation of a magnetic line switch *(Photo courtesy of General Electric Company)*

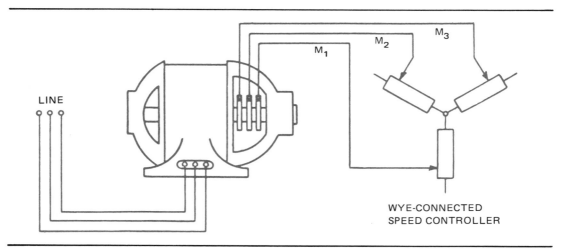

Fig. 11-2 Connections of a faceplate speed controller to a wound-rotor motor

speed controller box. The resistance value of each resistor section is varied by a contact arm. For the faceplate controller shown in figure 11-1, note that the three contact arms are connected at a common point at the center. A handle attached to one of the arms moves all arms simultaneously. These arms are spaced 120 mechanical degrees apart so that equal amounts of resistance can be cut in or out of each tapped resistor. This faceplate manual speed controller is connected in a three-phase wye arrangement.

Figure 11-2 shows the connections between a faceplate speed controller and a wound-rotor induction motor. As the three-phase, wound-rotor motor is started, all of the resistance in the speed controller is inserted in the rotor circuit. The stator circuit is connected across the three-phase line voltage by an across-the-line motor starter switch

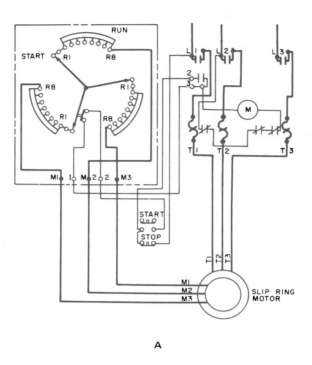

A

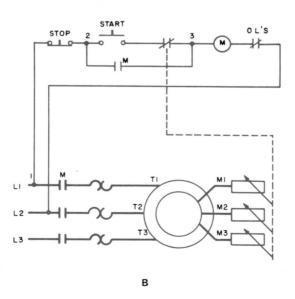

B

Fig. 11-3 Faceplate controller with a protective starting device A) Wiring diagram of a manual speed regulator interlocked with a magnetic starter B) Elementary diagram of figure 11-3A (From Alerich, *Electric Motor Control,* **copyright 1983 by Delmar Publishers Inc.)**

controlled from a pushbutton station. Because the maximum value of resistance is inserted in the rotor circuit at startup, the starting surge of current is limited. As a result, the starting torque is improved. After the motor has started, resistance is cut out of the rotor circuit using the speed controller until the desired speed is obtained.

Faceplate Controller with Protective Starting Device

If a wound-rotor induction motor is started with all of the speed controller resistance cut out of the rotor circuit, the starting surge of current to the stator windings will be high and the starting torque developed by the motor will be poor.

A protective starting device is used with a faceplate controller to insure that the motor is started with all of the resistance inserted (figure 11-3). The motor can be started only when the arms of this special type of controller are in the slow position with all of the resistance inserted in the rotor circuit.

A speed controller of this type has a pair of contacts which are closed when the three movable arms of the controller are in the slow position. These contacts are in series with the normally open start pushbutton. When the start pushbutton is pressed, the coil of the magnetic across-the-line starter switch is energized. The starter contacts close and the rated three-phase line voltage is applied to the stator windings. As the motor accelerates, the movable arms of the speed controller can be adjusted to obtain the desired speed. As the arms are moved from the slow position, the faceplate controller circuit contacts open. However, the contacts of the across-the-line magnetic motor switch are closed. Since one pair of these contacts acts as a sealing circuit around the normally open pushbutton and the open contacts of the circuit on the faceplate controller, the main starter coil remains energized and the motor continues to operate.

The motor cannot be started if the speed controller is not in the slow position. This is due to the fact that the control circuit to the coil of the magnetic motor switch is open because the contacts of the faceplate controller are open. Therefore, the motor switch will not operate when the start pushbutton is closed. To start the motor, the adjustable arms of the faceplate controller must be in the start position.

DRUM CONTROLLER

The drum controller, figure 11-4, is another type of manual speed controller which can be used with wound-rotor induction motors.

A drum controller consists basically of a case, contact fingers, the cylinder assembly, and external resistors. The case consists of a back piece and end pieces made of plate metal and a cover made of sheet metal. The cover fits over the end pieces and is removed when maintenance or repair work is required. The cover may be provided with a rubber gasket to make it dustproof. The wiring is brought to the controller through bushed holes or condulet fittings either in the back plate or the bottom end piece of the case.

The contact fingers of the drum controller are stationary contacts. The three wires from the rotor slip rings are connected to these contacts, as are the wires from the externally mounted grid resistors. Each contact finger is made of brass or steel and has a copper tip. Each finger is mounted and pivoted so that adjustments can be made on a spring to obtain the proper contact tension.

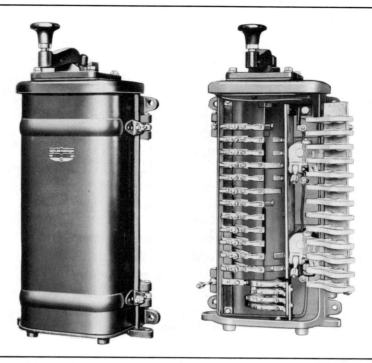

Fig. 11-4 Drum controller used for switching resistance values *(Photo courtesy of Cutler-Hammer Inc.)*

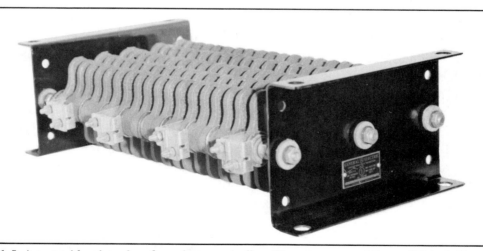

Fig. 11-5 A cast-grid resistor box for a drum controller *(Photo courtesy of General Electric Company)*

A drum controller also contains a vertical cylinder mounted on an insulated shaft. This cylinder provides the moving contacts that make and break connections to the various fixed contact fingers. The contacts on the cylinder are made of rolled copper segments which are moved by a handle located at the top of the speed controller.

The controller resistors usually are mounted outside and behind the controller case. The resistor units are cast from iron or a metal alloy, figure 11-5. Connecting wires from taps on the resistors terminate at contact fingers in the drum controller.

Operation of a Drum Controller

When the controller handle is in the slow position, the maximum resistance of the controller is inserted in the three phases of the rotor circuit. (The resistors of the speed controller are connected in wye.) As the controller handle is moved toward the run position, the copper contacts on the cylinder assembly make contact with various stationary fingers and sections of the resistance are cut out of the rotor circuit. When the handle is in the run position, all of the resistance of the speed controller is cut out of the rotor circuit. As a result, the motor operates at its rated speed.

Figure 11-6 is a diagram of the internal connections of a drum-type manual speed controller. This circuit indicates that in the run position, the leads from the rotor slip rings are connected together and the resistance of the controller is cut out of the rotor circuit.

The stator circuit is connected directly across the three-phase line by an across-the-line motor switch controlled from a pushbutton station. Some drum controllers may have a small pair of main relay coil contacts which are closed only when the speed controller is in the start position. These contacts are in series with the normally open start pushbutton. If an attempt is made to start the motor with the speed controller handle not in the slow position, the motor will not start. If the speed controller handle is returned to the slow position and the start pushbutton pressed, the motor will accelerate slowly. The operator can then adjust the speed of the motor to the desired value. After the motor has started, the coil contacts and the normally open start pushbutton is shunted out by the sealing contactors in the magnetic motor switch.

NATIONAL ELECTRICAL CODE REGULATIONS

The National Electrical Code in *Section 430-23* requires for continuous duty that the conductors from the rotor slip rings to the speed controller have an ampacity (current-carrying capacity) not less than 125 percent of the full-load current rating of the rotor, secondary circuit.

The Code lists several special industrial applications for which percentage values of the full load current other than 125 percent are allowed for the determination of wire size. As in the previous paragraph, for motors used in special industrial applications, the Code permits the application of other percentage values to the full-load current rating of the rotor to determine the rotor circuit wire size.

Resistor Duty Classification	Ampacity of Wire in Percent of Full-load Rotor Current
Light starting	35
Heavy starting	45
Extra heavy starting	55
Light intermittent	65
Medium intermittent	75
Heavy intermittent	85
Continuous	110

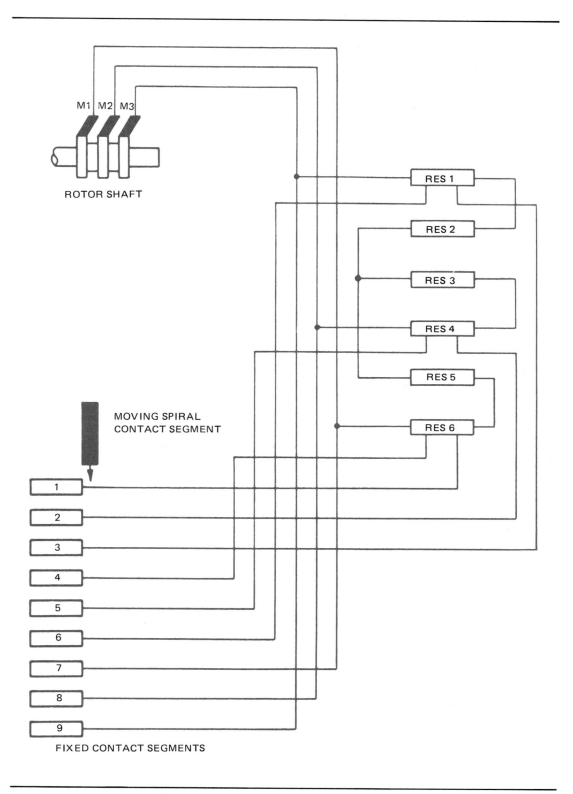

Fig. 11-6 Connections of a drum-type manual controller

If the speed controller resistors are separated from the speed controller, conductors are required to feed from the connection points on the resistors to the contact fingers. The Code also gives specific information on the percentage values of full-load current to be used to determine the size of these conductors.

The fusible starting protection for a wound-rotor induction motor is to be not more than 150 percent of the full-load motor current rating (NEC *Table 430-152*). This starting protection usually consists of fuses located in the motor disconnect switch. Wire ampacity is no less than 125 percent of the motor full load current (NEC *Section 430-22*).

Running overload protection is to be provided for wound-rotor induction motors. For a motor rated at more than one hp and marked to have a temperature rise of not more than 40 degrees Celsius, the running overload protection shall be rated at not more than 125 percent of the motor full-load current rating [NEC *Section 430-32(A)*]. The running overload protection usually consists of thermal overload units located in the magnetic motor starter.

The Code also states that the secondary circuits of wound-rotor induction motors, including conductors, controller, and resistors, shall be considered to be protected by the running overload devices provided in the primary circuit. [See NEC *Section 430-32(d)*.]

The phrase *primary circuit*, when referring to wound-rotor induction motors, means the stator winding. The phrase *secondary circuit*, as applied to wound-rotor induction motors, means the rotor circuit.

TERMINAL MARKINGS

The stator leads of a three-phase, wound-rotor induction motor are marked T_1, T_2, and T_3. Note that this is the marking system used with three-phase, squirrel-cage induction motors.

The rotor leads of a wound-rotor motor are marked M_1, M_2, and M_3. The M_1 lead connects to the slip ring nearest the bearing housing, lead M_2 connects to the middle slip ring, and lead M_3 connects to the slip ring nearest the rotor windings.

ACHIEVEMENT REVIEW

A.   Give complete answers to the following questions.

1. Give three reasons why speed controllers are used with wound-rotor induction motors. _____

    _____

    _____

    _____

2. List the two basic types of manual speed controllers used with wound-rotor induction motors. _____

    _____

    _____

3. A 230-volt, three-phase, 10-horsepower wound-rotor induction motor has a full-load stator current rating of 28 amperes and a full-load rotor current rating of 40 amperes per terminal. Determine the fuse size required for starting overload protection for this induction motor. _____

    _____

    _____

    _____

4. Determine the size of the thermal overload units required for the induction motor given in question 2. _____

    _____

    _____

    _____

5. What size wire, Type THHN, should be used for the stator circuit of the motor given in question 2? _____

    _____

    _____

    _____

6. What size wire, Type THHN, should be used for the connections between the slip rings and the speed controller for the motor in question 3? _____

    _____

    _____

    _____

7. What size wire, Type THHN, should be used between the external resistors and the speed controller in the motor in question 3 if the duty classification is "Light Intermittent Duty"? Check the Code before answering this problem.

    _____

    _____

    _____

    _____

8. Draw the wiring connection diagram of a wound-rotor induction motor which is started by means of an across-the-line magnetic motor starter switch controlled from a pushbutton station. Include a wye-connected speed controller in the rotor circuit.

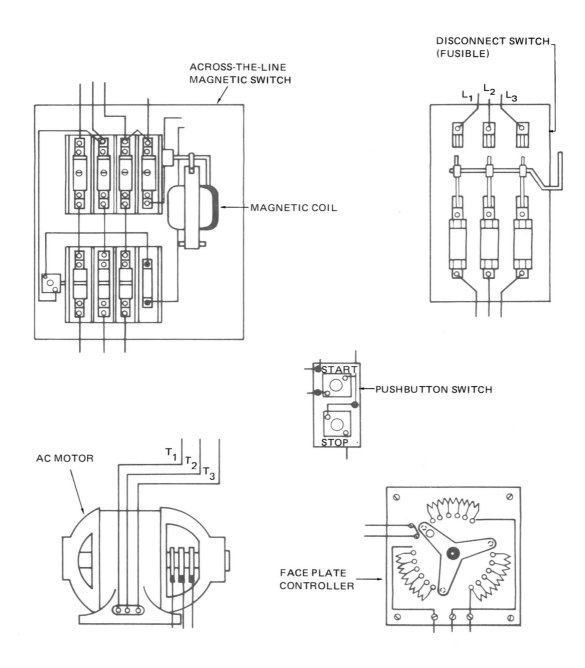

9. What precaution is provided on some speed controllers to prevent the starting of the motor when the speed controller is left in the run position? _____

_____

_____

_____

B. Select the correct answer for each of the following statements and place the corresponding letter in the space provided.

10. What type of speed controller is used with a wound-rotor motor? _____
    a. Primary resistor   c. Autotransformer
    b. Secondary resistor   d. Reactor

11. When starting a wound-rotor motor, full resistance is inserted in the secondary circuit and a magnetic starter applies full line voltage to
    a. the stator windings.   c. the starting resistors.
    b. the slip rings.   d. the primary resistors.

12. The rotor winding leads of a wound-rotor motor are brought out to _____
    a. two slip rings.   c. four slip rings.
    b. three slip rings.   d. a centrifugal switch.

13. The stator circuit of a wound-rotor motor is sometimes called the _____
    a. primary circuit.   c. rotor.
    b. secondary circuit.   d. frame.

14. The terminal lead connected to the slip ring nearest the bearing housing is _____
    a. T_1.   c. M_3.
    b. T_3.   d. M_1.

15. A drum controller is used for _____
    a. reversing the stator.
    b. switching resistances in the secondary.
    c. braking.
    d. reversing the rotor.

16. For best starting results, apply _____
    a. minimum voltage to the stator, with maximum resistance in the rotor.
    b. full voltage to the stator, with maximum resistance in the rotor.
    c. full voltage to the rotor, with maximum resistance in the stator.
    d. full voltage to the stator, with minimum resistance in the rotor.

THE SYNCHRONOUS MOTOR

OBJECTIVES

After studying this unit, the student will be able to

- list the basic parts in the construction of a synchronous motor.
- define and describe an amortisseur winding.
- describe the basic operation of a synchronous motor.
- describe how the power factor of a synchronous motor is affected by an under-excited dc field, a normally excited dc field, and an overexcited dc field.
- list at least three industrial applications of the synchronous motor.

The *synchronous motor,* figure 12-1, is a three-phase ac motor which operates at a constant speed from a no-load condition to full load. This type of motor has a revolving field which is separately excited from a direct-current source. In this respect, it is similar to a three-phase ac generator. If the dc field excitation is changed, the power factor of a synchronous motor can be varied over a wide range of lagging and leading values.

Fig. 12-1 Synchronous motor with direct-connected exciter *(Photo courtesy of General Electric Company)*

The synchronous motor is used in many industrial applications because of its fixed speed characteristic over the range from no load to full load. This type of motor also is used to correct or improve the power factor of three-phase ac industrial circuits, thereby reducing operating costs.

CONSTRUCTION DETAILS

A three-phase synchronous motor basically consists of a stator core with a three-phase winding (similar to an induction motor) a revolving dc field with an auxiliary or amortisseur winding and slip rings, brushes and brush holders, and two end shields housing the bearings that support the rotor shaft. An *amortisseur winding* (figure 12-2) consists of copper bars embedded in the cores of the poles. The copper bars of this special type of "squirrel-cage winding" are welded to end rings on each side of the rotor.

Both the stator winding and the core of a synchronous motor are similar to those of the three-phase, squirrel-cage induction motor and the wound-rotor induction motor. The leads for the stator winding are marked T_1, T_2, and T_3 and terminate in an outlet box mounted on the side of the motor frame.

The rotor of the synchronous motor has salient field poles. The field coils are connected in series for alternate polarity. The number of rotor field poles must equal the number of stator field poles. The field circuit leads are brought out to two slip rings mounted on the rotor shaft for brush-type motors. Carbon brushes mounted in brush holders make contact with the two slip rings. The terminals of the field circuit are brought out from the brush holders to a second terminal box mounted on the frame of the motor. The leads for the field circuit are marked F_1 and F_2. A squirrel-cage, or amortisseur, winding is provided for starting because the synchronous motor is not self-starting without this feature. The rotor shown in figure 12-2 has salient poles and an amortisseur winding.

AMORTISSEUR
(SQUIRREL-CAGE
WINDING)

SLIP RINGS

SALIENT POLES

Fig. 12-2 A synchronous motor rotor with amortisseur winding *(Photo courtesy of General Electric Company)*

Two end shields are provided on a synchronous motor. One of the end shields is larger than the second shield because it houses the dc brush holder assembly and slip rings. Either sleeve bearings or ball-bearing units are used to support the rotor shaft. The bearings also are housed in the end shields of the motor.

PRINCIPLE OF OPERATION

When the rated three-phase voltage is applied to the stator windings, a rotating magnetic field is developed. This field travels at the synchronous speed. As stated in previous units, the synchronous speed of the magnetic field depends on the frequency of the three-phase voltage and the number of stator poles. The following formula is used to determine the synchronous speed.

$$\text{Synchronous speed} = \frac{120 \times \text{frequency}}{\text{number of poles}}$$

$$S = \frac{120 \times f}{p}$$

The magnetic field which is developed by the stator windings travels at synchronous speed and cuts across the squirrel-cage winding of the rotor. Both voltage and current are induced in the bars of the rotor winding. The resulting magnetic field of the amortisseur (squirrel-cage) winding reacts with the stator field to create a torque which causes the rotor to turn.

The rotation of the rotor will increase in speed to a point slightly below the synchronous speed of the stator, about 92 percent to 97 percent of the motor rated speed. There is a small slip in the speed of the rotor behind the speed of the magnetic field set up by the stator. In other words, the motor is started as a squirrel-cage induction motor.

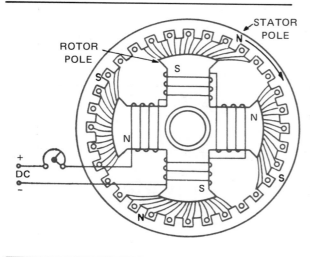

Fig. 12-3 Diagram to show the principle of operation of a synchronous motor

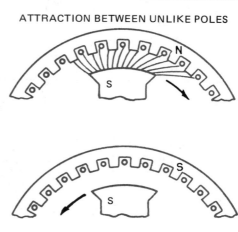

ATTRACTION BETWEEN UNLIKE POLES

REPULSION BETWEEN LIKE POLES

Fig. 12-4 Starting of synchronous motors

The field circuit is now excited from an outside source of direct current and fixed magnetic poles are set up in the rotor field cores. The magnetic poles of the rotor are attracted to unlike magnetic poles of the magnetic field set up by the stator.

Figures 12-3 and 12-4 show how the rotor field poles lock with unlike poles of the stator field. Once the field poles are locked, the rotor speed becomes the same as the speed of the magnetic field set up by the stator windings. In other words, the speed of the rotor is now equal to the synchronous speed.

Remember that a synchronous motor must always be started as a three-phase, squirrel-cage induction motor with the dc field excitation disconeected. The dc field circuit is added only after the rotor accelerates to a value near the synchronous speed. The motor then will operate as a synchronous motor, locked in step with the stator rotating field.

If an attempt is made to start a three-phase synchronous motor by first energizing the dc field circuit and then applying the three-phase voltage to the stator windings, the motor will not start since the net torque is zero. At the instant the three-phase voltage is applied to the stator windings, the magnetic field set up by the stator current turns at the synchronous speed. The rotor, with its magnetic poles of fixed polarity, is attracted first by an unlike stator pole and attempts to turn in that direction. However, before the rotor can turn, another stator pole of opposite polarity moves into position and the rotor then attempts to turn in the opposite direction. Because of this action of the alternating poles, the net torque is zero and the motor does not start.

Direct-Current Field Excitation

In the early models, the field circuit is excited from an external direct-current source. A dc generator may be coupled to the motor shaft to supply the dc excitation current.

Figure 12-5 shows the connections for a synchronous motor. A field rheostat in the separately excited field circuit varies the current in the field circuit. Changes in the field current affect the strength of the magnetic field developed by the revolving rotor. Variations in the rotor field strength do not affect the motor which continues to operate at a constant speed. However, changes in the dc field excitation do change the power factor of a synchronous motor.

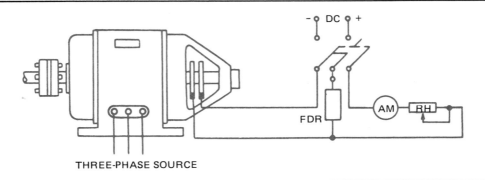

THREE-PHASE SOURCE

Fig. 12-5 External connections for a synchronous motor

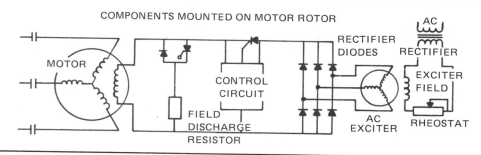

Fig. 12-6 Simplified circuit for a brushless synchronous motor

Brushless Solid-State Excitation

An improvement in synchronous motor excitation is the development of the brushless dc exciter. The commutator of a conventional direct-connected exciter is replaced with a three-phase, bridge-type, solid-state rectifier. The dc output is then fed directly to the motor field winding. Simplified circuitry is shown in figure 12-6. A stationary field ring for the ac exciter receives dc from a small rectifier in the motor control cabinet. This rectifier is powered from the ac source. The exciter dc field is also adjustable. Rectifier solid-state diodes change the exciter ac output to dc. This dc is the source of excitation for the rotor field poles. Silicon-controlled rectifiers, activated by the solid-state field control circuit, replace electromechanical relays and the contactors of the conventional brush-type synchronous motor.

The field discharge resistor is inserted during motor starting. At motor synchronizing pull-in speed, the field discharge circuit is automatically opened and dc excitation is applied to the rotor field pole windings. Excitation is automatically removed if the motor pulls out of step (synchronization) due to an overload or a voltage failure. The

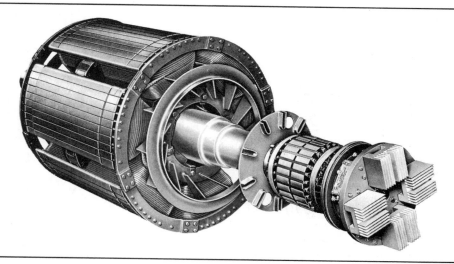

Fig. 12-7 Rotor of a brushless synchronous motor *(Photo courtesy of Electric Machinery, Turbodyne Division, Dresser Industries Inc.)*

stator of a brushless motor is similar to that of a brush-type motor. The rotor differs, however, as shown in figure 12-7. Mounted on the rotor shaft are the armature of the ac exciter, the ac output of which is rectified to dc by the silicon diodes. Brush and commutator problems are eliminated with this system.

Power Factor

A poor lagging power factor results when the field current is decreased below normal by inserting all of the resistance of the rheostat in the field circuit. The three-phase ac circuit to the stator supplies a magnetizing current which helps strengthen the weak dc field. This magnetizing component of current lags the voltage by 90 electrical degrees. Since the magnetizing component of current becomes a large part of the total current input, a low lagging power factor results.

If a weak dc field is strengthened, the power factor improves. As a result, the three-phase ac circuit to the stator supplies less magnetizing current. The magnetizing component of current becomes a smaller part of the total current input to the stator winding, and the power factor increases. If the field strength is increased sufficiently, the

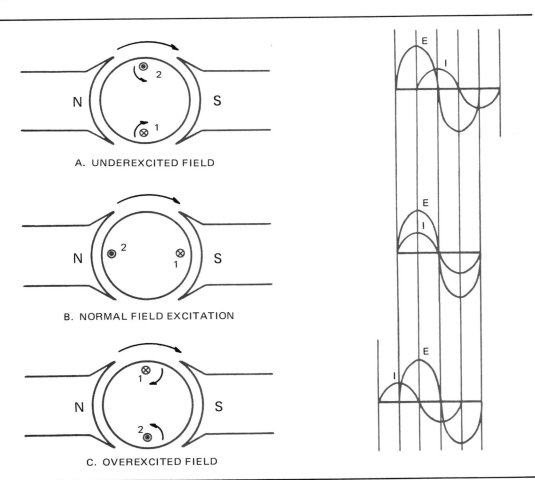

A. UNDEREXCITED FIELD

B. NORMAL FIELD EXCITATION

C. OVEREXCITED FIELD

Fig. 12-8 Field excitation in a synchronous motor

power factor increases to unity or 100 percent. When a power factor value of unity is reached, the three-phase ac circuit does not supply any current and the dc field circuit supplies all of the current necessary to maintain a strong rotor field. The value of dc field excitation required to achieve unity power factor is called *normal field excitation.*

If the magnetic field of the rotor is strengthened further by increasing the dc field current above the normal excitation value, the power factor decreases. However, the power factor is leading when the dc field is overexcited. The three-phase ac circuit feeding the stator winding delivers a demagnetizing component of current which opposes the too strong rotor field. This action results in a weakening of the rotor field to its normal magnetic strength.

The diagrams in figure 12-8 show how the dc field is aided or opposed by the magnetic field set up by the ac windings. It is assumed in figure 12-8 that the dc field is stationary and a revolving armature is connected to the ac source. The student should keep in mind the fact that most synchronous motors have stationary ac windings and a revolving dc field. For either case, however, the principle of operation is the same.

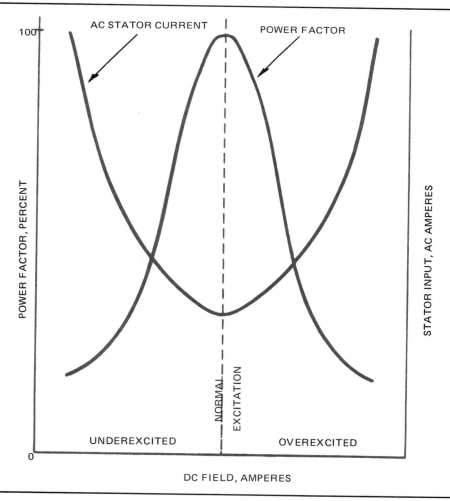

Fig. 12-9 Characteristic operating curves for synchronous motors

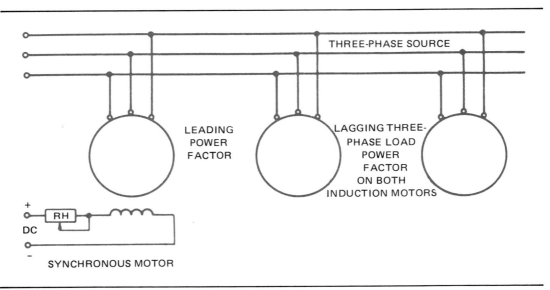

Fig. 12-10 Synchronous motor used to correct power factor

Figure 12-9 shows two characteristic operating curves for a three-phase synchronous motor. With normal full field excitation, the power factor has a peak value of unity or 100 percent and the ac stator current is at its lowest value. As the dc field current is decreased in value, the power factor decreases in the lag quadrant and there is a resulting rapid rise in the ac stator current. If the dc field current is increased above the normal field excitation value, the power factor decreases in the lead quadrant and a rapid rise in the ac stator current results.

It has been shown that a synchronous motor operated with an overexcited dc field has a leading power factor. For this reason, a three-phase synchronous motor often is connected to a three-phase industrial feeder circuit having a low lagging power factor (figure 12-10). In other words, the synchronous motor with an overexcited dc field will correct the power factor of the industrial feeder circuit.

In figure 12-10, two induction motors with lagging power factors are connected to an industrial feeder circuit. The synchronous motor connected to the same feeder is operated with an overexcited dc field. Since the synchronous motor can be adjusted so that the resulting power factor is leading, the power factor of the industrial feeder can be corrected until it reaches a value near unity or 100 percent.

Reversing Rotation

The direction of rotation of a synchronous motor is reversed by interchanging any two of the three line leads feeding the stator winding. The electrician must remember that the direction of rotation of the motor does not change if the two conductors of the dc source are interchanged.

INDUSTRIAL APPLICATIONS

The three-phase synchronous motor is used when a prime mover having a constant speed from a no-load condition to full load is required, such as fans, air compressors, and pumps. The synchronous motor is used in some industrial applications to drive a mechanical load and also correct power factor. In some applications, this type of motor is used only to correct the power factor of an industrial power system. When the synchronous motor is used only to correct power factor and does not drive any mechanical load, it serves the same purpose as a bank of capacitors used for power factor correction. Therefore, in such an installation the motor is called a *synchronous capacitor.*

Three-phase synchronous motors, up to a rating of 10 horsepower, usually are started directly across the rated three-phase voltage. Synchronous motors of larger sizes are started through a starting compensator or an automatic starter. In this type of starting, the voltage applied to the motor terminals at the instant of start is about half the value of the rated line voltage and the starting surge of current is limited.

ACHIEVEMENT REVIEW

A.   Completely answer the following questions.

1. List the basic parts of a three-phase synchronous motor. _____

_____

_____

2. What is an amortisseur winding? _____

_____

_____

3. Explain the proper procedure to use in starting a synchronous motor. _____

_____

_____

_____

4. A three-phase synchronous motor with six stator poles and six rotor poles is operated from a three-phase, 60-hertz line of the correct voltage rating. Determine the speed of the motor. _____

_____

_____

5. How is a leading power factor obtained with a three-phase synchronous motor?

_____

_____

_____

6. What is the purpose of a rheostat in the separately excited dc field circuit of a synchronous motor? _____

_____

_____

7. How is the direction of rotation of a three-phase synchronous motor reversed?

   _____

   _____

   _____

8. State two important applications for three-phase synchronous motors.

   _____

   _____

   _____

9. What is a synchronous capacitor? _____

   _____

B. Select the correct answer for each of the following statements and place the corresponding letter in the space provided.

10. A synchronous motor must be started   _____
    a. with full dc in the field circuit.
    b. with weak dc in the field circuit.
    c. as an induction motor.
    d. when the power factor is low.

11. The speed of a synchronous motor   _____
    a. is constant from no load to full load.
    b. drops from no load to full load.
    c. increases from no load to full load.
    d. is variable from no load to full load.

12. A synchronous motor with an underexcited dc field has _____
    a. a leading power factor.
    b. a lagging power factor.
    c. less synchronous speed.
    d. no effect.

13. The power factor of a synchronous motor can be varied by changing the   _____
    a. brush polarity.
    b. phase rotation.
    c. speed of rotation.
    d. field excitation.

14. A synchronous motor running on the three-phase line voltage serves the same function of power factor correction as   _____
    a. a bank of resistors.
    b. a bank of capacitors.
    c. an induction motor.
    d. a wound-rotor motor.

CONTROLLERS FOR THREE-PHASE MOTORS

OBJECTIVES

After studying this unit, the student will be able to

- describe the basic sequence of actions of the following types of controllers when used to control three-phase ac induction motors:
  - jogging-type controller,
  - quick-stop ac controller (plugging),
  - dynamic braking controller,
  - resistance starter controller,
  - automatic autotransformer compensator,
  - automatic controller for wound-rotor induction motors,
  - wye-delta controller, and
  - automatic controller for synchronous motors.
- identify and use the various National Electrical Code sections pertaining to controllers and remote control circuits for motors.
- state why ac adjustable speed drives are used.
- list the types of adjustable speed drives.
- describe the operating principles of different ac adjustable speed drives.
- list the advantages and disadvantages of some units.

The industrial electrician is required to install, maintain, and repair automatic ac controllers which start up and provide speed control for squirrel-cage induction motors, wound-rotor induction motors, and synchronous motors.

MOTOR CONTROLLERS WITH JOGGING CAPABILITY

Many industrial processes require that the driven machines involved in the process be inched or moved small distances. Motor controllers designed to provide control for this type of operation are called jogging controllers. *Jogging* is defined as the quickly repeated closure of a controller circuit to start a motor from rest for the purpose of accomplishing small movements of the driven machine.

An across-the-line magnetic motor switch may be used to provide jogging control if the proper type of pushbutton station is used in the control circuit. Such a pushbutton station is called a *start-jog-stop station*.

Figure 13-1A is a diagram of the connections for a three-phase, squirrel-cage induction motor connected to a jogging-type, across-the-line motor starting switch.

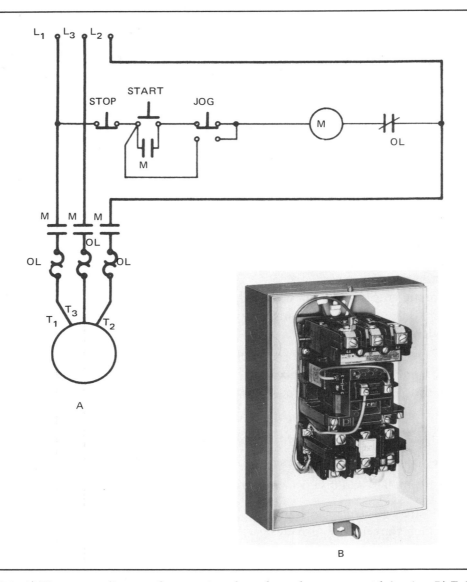

Fig. 13-1 A) Elementary diagram of connections for a three-phase motor with jogging **B)** Full voltage starter, size 1 (with cover removed) *(Photo courtesy of Allen-Bradley Company)*

Figure 13-1B shows the starter with the cover removed. Note in figure 13-1 that the connections and operation of the start and stop pushbuttons are the same as those of a standard pushbutton station with start and stop positions. The connections for the jog pushbutton, however, are more complex and should be studied in detail. When the jog pushbutton is pressed, coil **M** is energized, main contacts **M** close, and the motor starts turning. The small auxiliary contacts **M** also close, but do not function as a sealing circuit around the jog pushbutton because pushing the jog pushbutton also opens the sealing circuit. As a result, as soon as the jog pushbutton is released, coil **M** is de-energized and all **M** contacts open. Before the jog pushbutton returns to its normal position, the sealing contacts M open and thus the control circuit remains open. This control also can be used for standard start-stop operations. In summary then, repeated closures of the jog pushbutton start and stop the motor so that the driven machine can be inched or jogged to the desired position.

QUICK-STOP AC CONTROLLER (PLUGGING)

Some industrial applications require that three-phase motors be stopped quickly. If any two of the line leads feeding a three-phase motor are reversed, a counter torque is set up which brings the motor to a quick standstill before it begins to rotate in the reverse direction. If the circuit is interrupted at the instant the motor begins to turn in the opposite direction, the rotor will just start to turn in this direction and then will stop. This method of bringing a motor to a quick stop is called *plugging*. The motor controller required to provide this type of operation is an across-the-line magnetic motor starter with reversing control and a special plugging relay. The plugging relay is belt-driven from an auxiliary pulley on the motor shaft, or onto a through shaft motor.

The connections for a quick-stop ac controller are shown in figure 13-2A. The controller itself is shown in figure 13-2B. When the start pushbutton is pressed, relay coil F is energized. As a result, the small, normally closed F contacts open. These contacts are connected in series with the reverse coil, which locks out reverse operation. In addition, the other small, normally open F contacts close and maintain the start pushbutton circuit. When the start button is released, the circuit of coil F is maintained through the sealing circuit, main contacts F close, and the rated three-phase voltage is applied to the motor terminals. Then, the motor comes up to speed and contacts PR of the plugging relay close.

When the stop pushbutton is pressed, the F relay coil is deenergized. As a result of this action, the main F contacts for the motor circuit open and disconnect the motor from the three-phase source. In addition, the small F sealing contacts open, deenergizing the holding or sealing circuit around the start pushbutton. Finally, the small F contacts in series with the reverse relay close to their normal position and the reverse relay coil is energized.

The main R contacts now close to reconnect the three-phase line voltage to the motor terminals. The connections of two line leads are interchanged. The resulting reversing countertorque developed in the motor brings it to a quick stop. At the moment the motor begins to turn in the reverse direction, contacts PR open due to the mechanical action of the PR relay unit. Coil R is deenergized and the R contacts open and interrupt the power supply to the motor. Since the motor is just beginning to turn in the opposite direction, it comes to a standstill. The motor supplies the mechanical power to drive a disk which causes contacts PR to close when the motor is in operation.

DYNAMIC BRAKING WITH INDUCTION MOTORS

It should be recalled from the study of dc controllers that dynamic braking is a method used to help bring a motor to a quicker stop without the extensive use of friction brakes. In this application, dynamic braking means that the motor involved is

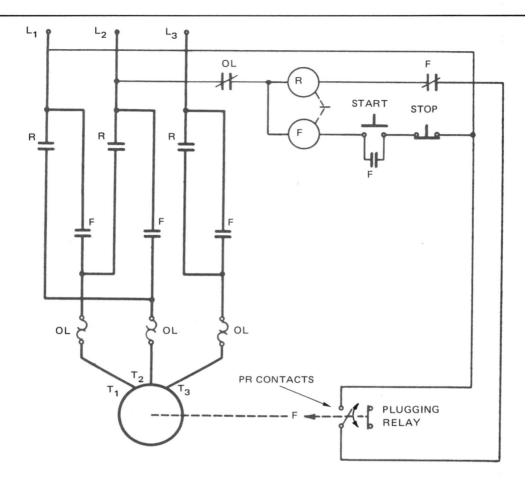

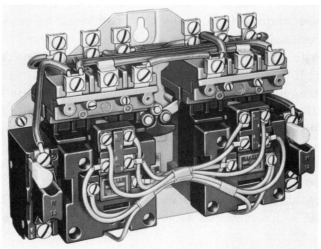

Fig. 13-2   A) Elementary circuit with a plugging relay   **B)** Ac full voltage reversing starter, size 1
(Photo courtesy of Allen-Bradley Company)

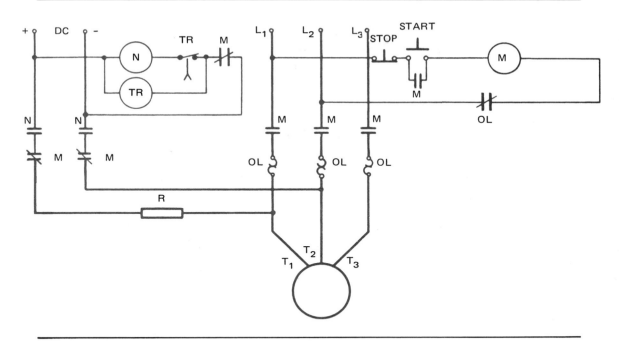

Fig. 13-3 A circuit for dynamic braking of an ac motor

used as a generator. An energy dissipating resistance is connected across the terminals of the motor after it is disconnected from the line.

Dynamic braking also can be applied to induction motors. When the stop pushbutton is pressed, the motor is disconnected from the three-phase source and the stator windings are excited by a dc source. A stationary magnetic field is developed by the direct current in the stator windings. As the squirrel-cage rotor revolves through this field, a high rotor current is created. This rotor current reacts with the stationary field of the stator to produce a countertorque that slows and stops the motor.

Figure 13-3 is a diagram of an ac motor installation with an across-the-line magnetic motor starter and dynamic braking capability.

When the start pushbutton is pressed, coil M is energized. At this instant, the main M contacts close and connect the motor terminals to the three-phase source, and the auxiliary, normally open M contacts close and provide a maintaining circuit around the start pushbutton. When relay coil M is energized, the normally closed contacts M in the dc control circuit open, with the result that both the main dc relay coil N and the time-delay relay coil TR are deenergized and interlocks in the dc circuit open. The three-phase voltage applied to the motor terminals causes the motor to accelerate to the rated speed.

When the stop pushbutton is pressed, coil M is deenergized. At this moment, a number of actions occur: 1) the main contacts M open and disconnect the motor from the three-phase source; 2) the auxiliary M contacts open (these contacts act as a maintaining circuit); 3) protective interlocks M in the dc circuit close; and 4) the auxiliary, normally closed contacts M in the dc control circuit close and energize the time-delay

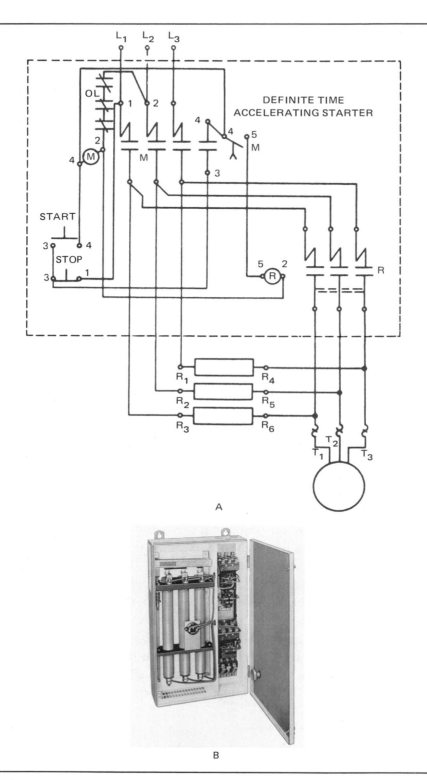

A

B

Fig. 13-4   A) Definite time acceleration with resistance starting   B) Two-point primary resistance starter rated at 25 hp, 600 V *(Photo courtesy of Allen-Bradley Company)*

relay and the main relay coil N. Energizing relay coil N causes the closing of contacts N so that dc voltage is connected on the ac windings through a current-limiting resistance. As a result, the motor comes to a quick stop. Following a definite period after the motor has stopped, measured in seconds, relay coil TR operates and opens contacts TR to cause coil N to become deenergized. Thus, contacts N open and disconnect the motor windings from the dc source. The controller now is ready for the next starting cycle.

RESISTANCE STARTER CONTROLLER

When a squirrel-cage induction motor is connected directly across the rated line voltage, the starting current may be 300 percent to 600 percent of the rated current of the motor. In large motors, this high current may cause serious voltage regulation problems and overloading of industrial power feeders.

The starting current of a squirrel-cage induction motor can be reduced by using a resistance starter controller. This type of controller inserts equal resistance values in each line wire at the instant the motor is started. After the motor accelerates to a value near its rated speed, the resistances are cut out of the circuit and full line voltage is applied to the motor terminals.

Figure 13-4A is a diagram of the circuit connections for a resistance starter. A photo of this starter is figure 13-4B. When the start button is pressed, main relay coil M is energized. The main contacts close and connect the motor to the three-phase source through the three resistors (R). The circuit for coil M is maintained through the small auxiliary contacts (3 and 4) which act as a sealing circuit around the normally open start pushbutton. When the main contacts of relay coil M are closed, a mechanical device, called a *definite time relay,* is started. After a predetermined time elapses, the definite time contacts close and energize coil R. Coil R causes three sets of contacts to close and shunt out the three resistors. Thus, the motor is connected directly across the rated line voltage with no interruption of the power line (closed transition).

When the stop button is pressed, the circuits of both coil M and coil R are opened. This causes the opening of the main contacts, the sealing contacts, and the contacts which shunt the series resistors. As a result, the motor is disconnected from the three-phase source.

The starting current in the resistance starter causes a relatively high voltage drop in the three resistors. Because of this, the voltage across the motor terminals at start is low. As the motor accelerates, the current decreases, the voltage drop across the three resistors decreases, and the terminal voltage of the motor increases gradually. A smooth acceleration is obtained because of this gradual increase in the terminal voltage. However, it may be unwise to select resistance starting for many starting tasks because of the energy dissipated in the starting resistors.

The National Electrical Code provides guidelines on the selection of the correct fuse sizes for starting protection on a branch motor circuit containing a squirrel-cage induction motor with a resistance starter. In addition, the Code specifies the running overload protection required, and the wire sizes required on the branch motor circuit. (See NEC *Article 430.*)

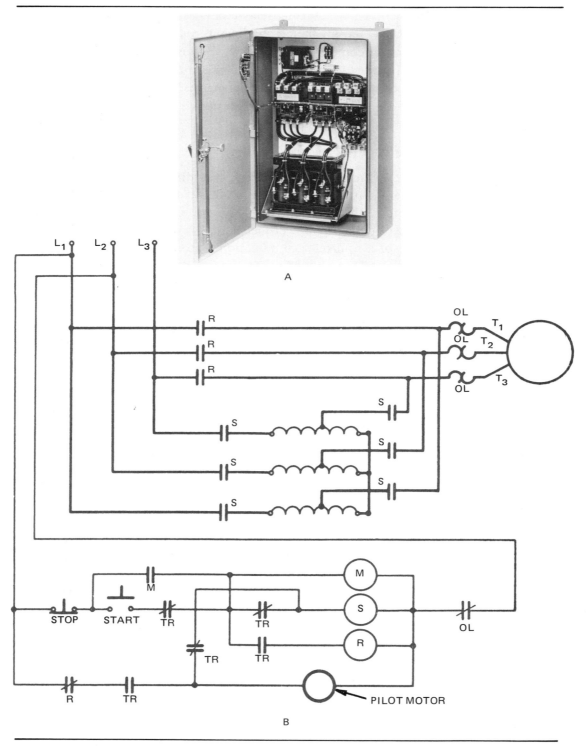

Fig. 13-5 **A)** Reduced voltage starter, autotransformer type, with a pneumatic timer *(Courtesy of Allen-Bradley Company)* **B)** Elementary diagram of an autotransformer compensator for starting an induction motor

AUTOMATIC AUTOTRANSFORMER COMPENSATOR

The automatic autotransformer (figure 13-5A) compensator basically operates in the same manner as the manual starting compensator. The automatic compensator has the advantage that it can be pushbutton-controlled from a convenient location. Figure 13-5B is a typical schematic wiring diagram for an automatic autotransformer compensator.

If the start button is pressed, a circuit is established from line 1 through the following devices to line 2: the normally closed stop button, the start button, contacts TR (timing relay) to relay coil S (start coil), and the normally closed overload contacts. Coil M is energized, closing the normally open small contacts M to provide the maintaining circuit.

When coil S is energized, all contacts marked S close. The three autotransformers are connected in wye across the three-phase line and supply reduced voltage to the motor. The motor begins to accelerate to a value near the rated speed.

As shown in figure 13-5B, a second circuit is established through the pilot motor by contacts TR. The pilot motor begins operating as soon as it is energized. After a definite timed period (figure 13-6) the pilot motor mechanically actuates all TR contacts. Coil S is deenergized as a result. This coil was used to start the three-phase

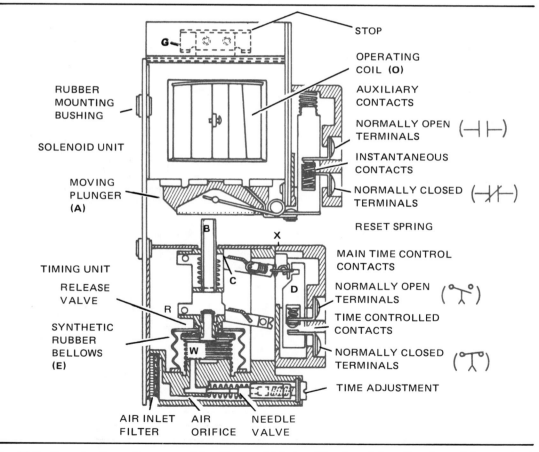

Fig. 13-6 Cross section of an ac on-delay timer, which provides time delay after the coil is energized. It is shown with the coil energized and the timer timed out. Schematic wiring symbols are shown (de-energized positions) for various portions of the timer.

motor on reduced voltages. At the same time, coil R (run coil) is energized and closes the running contacts to apply the rated three-phase voltage to the motor terminals.

When contacts TR open, the pilot motor circuit is opened, and the pilot motor stops. At this time, the motor is operating on full voltage. If the stop button is pressed, the holding coil circuit for coils M and R opens. As a result, the R contacts to the motor open and the motor stops. When coil R is deenergized, the normally closed R contacts close the pilot motor circuit. The pilot motor runs to set all TR contacts for the next starting cycle.

The National Electrical Code rulings for starting and running protection also apply to motors operated with either a manually operated starting compensator or an automatic autotransformer compensator.

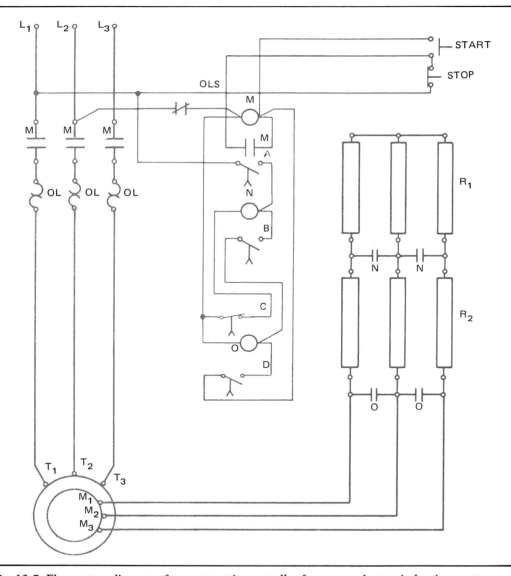

Fig. 13-7 Elementary diagram of an automatic controller for a wound-rotor induction motor

AUTOMATIC CONTROLLER FOR WOUND-ROTOR INDUCTION MOTORS

Manual speed controllers, such as the faceplate type or the drum type, are used to provide speed control for wound-rotor induction motors in industrial applications. If the resistance in the rotor circuit of a wound-rotor induction motor is to be used only on starting, then an automatic controller may be used, figure 13-7. In this case, resistors in the rotor circuit are automatically cut out by contactors arranged to operate in sequence at definite time intervals.

As shown in figure 13-7, when the start button is pressed, the main relay coil M is energized. The main contacts are closed to connect the stator circuit of the motor directly across the three-phase line voltage. All of the resistance of the controller is inserted in the secondary circuit of the motor as it begins to accelerate.

After the start button is released to its normally open position, the small auxiliary contacts M act as a maintaining circuit to keep the circuit of coil M closed. Contacts A are held open for a timed period (seconds) by a mechanical or electronic device (figure 13-8A, B, and C). When the A contacts close, coil N is energized through the normally closed contacts C, and all N contacts close to shunt out the R_1 resistors in the rotor circuit.

Contacts B also are held open for a definite number of seconds by a mechanical or electronic device. When B contacts close, coil O is energized, all O contacts close, and all resistance is cut out of the rotor circuit. At the same time, the C contacts open and deenergize coil N which then opens contacts N. The D contacts then close and maintain a closed circuit through coil O.

When the stop pushbutton is pressed, relay coil M is deenergized, and contacts M open to disconnect the motor from the line. Coil O also is deenergized and contacts O open, with the result that all of the resistance is inserted in the rotor circuit for the next starting cycle.

The National Electrical Code regulations for wire size, starting overload protection, and running overload protection also apply to both manual speed controllers and automatic controllers used with wound-rotor induction motors.

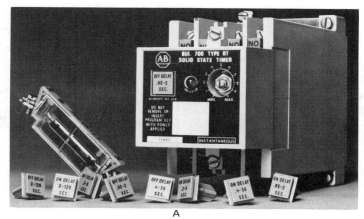

A

Fig. 13-8 Solid-state timing relays with different plug-in program keys *[(A) courtesy of Allen-Bradley Company, (B and C) courtesy of Square D Company)]* (continued)

1. STANDARD INDUSTRIAL CONTROL RELAY MOUNTING
2. REMOVABLE TIMER COVER PROTECTS TIME DELAY AND MODE SETTING
3. LED (LIGHT EMITTING DIODE) TIMING INDICATOR
4. CONVERTIBLE TIME DELAY MODE SHOWS THROUGH COVER
5. ONE N.O. AND ONE N.C. TIMED NEMA B150 HARD OUTPUT CONTACTS (10 AMPERE CONTINUOUS)
6. TERMINALS CLEARLY MARKED
7. FIVE TIMING RANGES FROM 0.05 SECONDS TO 10 HOURS
8. MARKING AREA
9. SELF-LIFTING PRESSURE WIRE CONNECTORS
10. OPTIONAL INSTANTANEOUS NEMA B150 HARD OUTPUT CONTACTS (10 AMPERE CONTINUOUS)

B

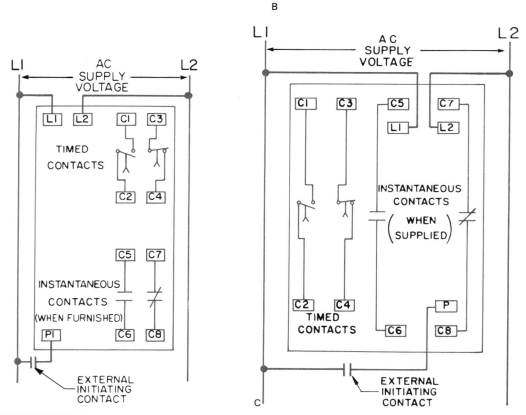

Fig. 13-8 Solid-state timing relays with different plug-in program keys *[(A) courtesy of Allen-Bradley Company, (B and C) courtesy of Square D Company)]* (continued)

WYE-DELTA CONTROLLER

Figure 13-9A shows a simple method by which a three-phase, delta-connected motor can be started on reduced voltage by connecting the stator windings of the motor in wye during the starting period. Figure 13-9B shows the actual starter. After the motor accelerates, the windings are reconnected in delta and placed directly across the rated three-phase voltage.

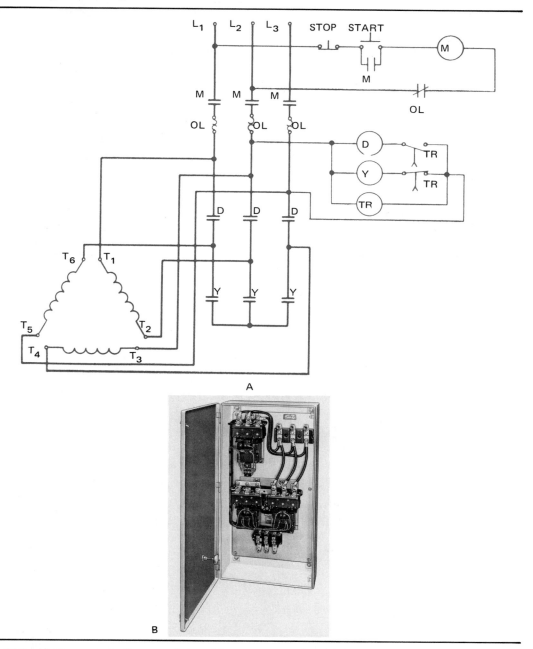

Fig. 13-9 A) Elementary diagram of wye-delta motor starting **B)** Star-delta starter rated at 75 hp, 240 V *(Photo courtesy of Allen-Bradley Company)*

When the start button is pressed, the main M contacts close, and relay coil Y and time-delay relay TR are energized. Coil Y causes contacts Y to close and the windings of the motor are connected in wye. If the line voltage is 230 volts, the voltage across each winding is:

$$\frac{230}{1.73} = 133 \text{ volts}$$

The voltage across each winding is only 58 percent of the line voltage when the windings are connected in wye at the start position.

After a definite period of time, the time-delay relay TR opens the circuit of relay coil Y and the Y contacts open.

Then, the time-delay relay TR closes the circuit of relay coil D. All D contacts are closed and the motor winding connections are changed from wye to delta. Full line voltage is applied to the motor windings and the motor operates at its rated speed.

Motors started by a wye-delta controller must have the leads of each phase winding brought out to the terminal connection box of the motor. In addition, the phase windings must be connected in delta for the normal running position. NOTE: The electrician should never attempt to operate a three-phase, wye-connected motor with this type of controller. This is due to the fact that there will be an excessive voltage applied to the motor windings in the run position when the windings are connected in delta by the controller.

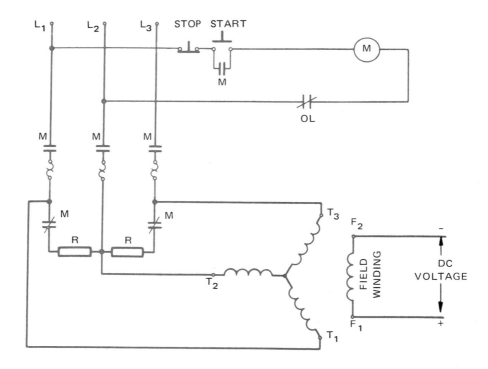

Fig. 13-10 Synchronous motor controller with dynamic braking

AUTOMATIC CONTROLLER FOR SYNCHRONOUS MOTORS

Synchronous motors may be started by means of an across-the-line magnetic motor starting switch, a manual starting compensator, or an automatic starting compensator. Dynamic braking may be provided by the controller.

Figure 13-10 is a diagram of the connections for a synchronous motor controller with dynamic braking. When the start button is pressed, main relay coil M is energized. The four normally open M contacts close and the two normally closed M contacts open. Three-phase voltage is applied to the motor terminals. When the motor accelerates to a speed near the synchronous speed, the dc field circuit is energized by secondary controls.

When the stop button is pressed, main relay coil M is deenergized. The M contacts open and disconnect the motor terminals from the three-phase line. The two normally closed M contacts reconnect the motor windings through the resistors and the dc field remains energized. As a result, the synchronous motor acts as an ac generator and delivers electrical energy to the two R resistors. The use of this type of controller results in a more rapid slowing of a synchronous motor.

The National Electrical Code provides guidelines for branch-circuit fuse protection and running overload protection for branch circuits feeding three-phase synchronous motors, and for allowable conductor sizes for branch circuits feeding synchronous motors. Local building and electrical code authorities should be consulted before installations are made with motors and controllers which do not comply with National Electrical Code rulings.

SOLID-STATE REDUCED VOLTAGE STARTERS

Solid-state devices and equipment are used for reduced voltage motor starting, electrical energy saving control circuits, variable speed drives, motor protection and other applications. A motor starter consists of a control circuit, a motor power circuit, and protective devices for the wiring and the motor. The functions of a starter are performed by contactors and overload relays in electromechanical motor starters. In solid-state starters, the control functions are performed by semiconductors. They are controlled by integrated circuits and microprocessors to provide the protective functions, operating instructions, and control.

Construction and Operation

The solid-state reduced-voltage starter provides a smooth, stepless acceleration of a three-phase induction motor. This is accomplished by gradually turning on six power SCRs (silicon controlled rectifiers). Two SCRs per phase are connected in a back-to-back or reverse parallel arrangement, figure 13-11. The SCRs are mounted on a heat (dissipating) sink to make up a power pole (phase). Each power pole contains the gate firing circuits as discussed in unit 7. An integrated thermal sensor is also provided to deenergize the starter if an over-temperature condition exists.

The firing circuitry on each power pole is controlled by a logic module. These modules monitor the starter for correct start up and operating conditions. Some motor starters provide a visual indicator of the starting condition through the use of light-emitting diodes (LEDs).

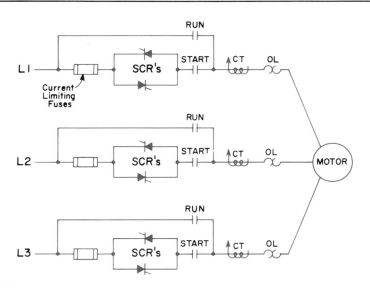

Fig. 13-11 Solid-state reduced voltage starter power circuit *(Photo courtesy of Allen-Bradley Company)*

The SCRs are connected back to back so that they may pass ac and control the amount of voltage. The current-limiting starter is a common type; it is designed to maintain the motor current at a constant level throughout the acceleration period. A current-limit potentiometer adjustment is provided to preset this current. A starter with current ramp acceleration is designed to begin acceleration at a low current level and then increase the current during the acceleration period.

As indicated in figure 13-11, this starter includes both *start* and *run* contactors. The start contacts are in series with the SCRs; the run contacts are in parallel with the combination of SCRs and start contacts. When the starter is energized, the start contacts close. The motor acceleration is then controlled by phasing-on the SCRs. When the motor reaches full speed, the run contacts close and the motor is connected directly across the lines (closed transition). At this point, the SCRs are turned off and the start contacts open. Under full speed running conditions, the SCRs are out of the circuit, eliminating SCR power losses during the run cycle. This feature saves energy; it also guards against possible damage due to overvoltage transients. With the starter in the deenergized position, all contacts are open, isolating

Fig. 13-12 Solid-state reduced voltage starter power circuit *(Photo courtesy of Allen-Bradley Company)*

the motor from the line. This open circuit condition protects against accidental motor rotation as a result of SCR misfiring and/or SCR damage caused by overvoltage transients. A solid-state reduced-voltage starter is shown in figure 13-12. Field connections are very similar to those for electromechanical starters.

Reduced Voltage Operation

To reduce the voltage applied to the motor in a solid-state starter, the SCRs can be turned on by the "gate" electrode for any desired part of each half cycle. Usually the SCRs turn off as the current wave reaches zero. They stay off until gated on again in the next half cycle. Some devices can vary the switching and timing. By switching the controlled current gating, the effective ac voltage can be varied to the motor. This voltage can be varied from zero to full voltage as required. The voltage is applied at some preset minimum value that can start the motor rotating. As the motor speed builds up, the SCR "on" time is gradually increased. The voltage is increased until the motor is placed across the line at full voltage. Mechanical shock is reduced and the current inrush can be regulated and controlled as desired (figure 13-13). The solid-state reduced-voltage starter can replace any of the electromechanical starters already described for reduced voltage starting.

CODE REFERENCES FOR MOTOR CONTROLLERS

The following sections of the National Electrical Code are concerned with motor controllers and remote control circuits.

1. *Sections 430-8* and *430-9* refer to the identification of motors and controllers with respect to controller nameplate ratings and terminal markings.

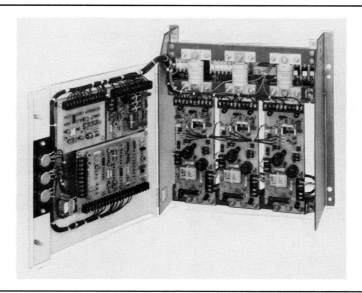

Fig. 13-13 **SCR controller section, the regulating part of the starter. The controller determines to what degree the SCRs should be phased on, thereby controlling the voltage applied to the motor** *(Photo courtesy of Allen-Bradley Company)*

2. *Section 430-C* is concerned with overload protection.
3. *Section 430-37* gives the number of running overcurrent relays required for different electrical systems.
4. *Section 430-D* is concerned with branch-circuit protection.
5. *Sections 430-71* to *430-74* are concerned with the control circuits of controllers.
6. *Sections 430-81* to *430-90* are concerned with controller installations.
7. *Sections 430-101* to *430-113* cover motor disconnecting means.

AC ADJUSTABLE SPEED DRIVES

Adjustable speed drives have a flexibility that is particularly useful in specialized applications. For this reason, these drives are widely used throughout industry for conveyors used to move materials, hoists, grinders, mixers, pumps, variable speed fans, saws, and crushers. The advantages of ac drives include the maximum utilization of the driven equipment, better coordination of production processes, and reduced wear on mechanical equipment.

The ac induction motor is the major converter of electrical energy into another usable form. About two-thirds of the electrical energy produced in the United States is delivered to motors.

Much of the power that is consumed by ac motors goes into the operation of fans, blowers, and pumps. It has been estimated that approximately 50% of the motors in use are for these types of loads. Such loads are particularly appropriate to look at for energy savings. Several alternate methods of control for fans and pumps have been developed and show energy savings over traditional methods of control.

Fans and pumps are designed to meet the maximum demand of the system in which they are installed. Often, however, the actual demand varies and may be much less than the design capacity. Such conditions are accommodated by adding outlet dampers to fans or throttling valves to pumps. These controls are effective and simple, but affect the efficiency of the system. Other forms of control have been developed to adapt machinery to varying demands. These controls do not decrease the efficiency of the system as much as the traditional methods of control. One of the newer methods is the direct variable speed control of the fan or pump. This method produces a more efficient means of flow control compared to the other existing methods.

In addition to a tangible reduction in the power required to operate equipment and machinery resulting from the use of adjustable speed drives, other benefits include extended bearing life and pump seal life.

WOUND ROTOR AC MOTORS

Wound rotor motor drives use a specially constructed ac motor to accomplish speed control. The windings of the motor rotor are brought out of the motor through slip rings on the rotor shaft. Figure 13-14 shows an elementary diagram of a wound rotor motor with an adjustable speed drive. These windings are connected to a controller which places variable resistors in series with the windings. The torque performance of the motor can be controlled using these variable resistors.

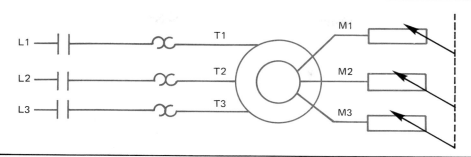

Fig. 13-14 Elementary diagram of an adjustable speed drive wound rotor motor

Wound rotor motors are most common in the larger sizes, in the range of 300 horsepower and above.

Features of Wound Rotor Motors

Wound rotor motors have the following advantages which make them suitable for a variety of applications:

- **Cost** — the initial cost is moderate for the high horsepower units.
- **Control** — not all the power need be controlled, resulting in a moderate size and simple controller.
- **Construction** — the simple construction of the motor and control lends itself to maintenance without the need for a high level of training.
- **High inertia loads** — the drive works well on high inertia loads.

Disadvantages of Wound Rotor Motors

- **Custom motor** — the motor has a rotor wound with wire, slip rings, and is not readily available.
- **Efficiency** — the drive does not maintain a high efficiency at low speeds.
- **Speed range** — the drive usually is limited to a speed range of two to one.

TYPES OF ADJUSTABLE SPEED DRIVES

Several types of variable speed drives can be used with wound rotor induction motors. These drives are eddy current (magnetic) drives, variable pitch drives, and adjustable frequency drives.

Eddy Current (Magnet) Drives

The eddy current drive couples the motor to the load magnetically (figure 13-15). The electromagnetic coupling is a simple way to obtain an adjustable output speed from the constant input speed of squirrel cage motors. There is no mechanical contact between the rotating members of the eddy current drive; thus, there is no wear. Torque is transmitted between the two rotating units by an electromagnetic reaction created by an energized rotating coil winding. The rotation of the ring with relation to the

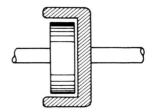

Fig. 13-15 Spider rotor coil magnet rotates within a steel drum (From Alerich, *Electric Motor Control,* copyright 1983 by Delmar Publishers Inc.)

electromagnet generates eddy currents and magnetic fields in the ring. Magnetic interaction between the two units transmits torque from the motor to the load. The slip between the motor and the load can be controlled continuously with great precision.

Torque can be controlled using a thyristor in an ac or dc circuit, or by using a rheostat to control the field through slip rings. When the eddy current drive responds to an input or command voltage, the speed of the driven machine changes. A further refinement can be obtained in automatic control to regulate and maintain the output speed. The magnetic drive can be used with nearly any type of actuating device or transducer that can provide an electrical signal. For example, the input can be provided by static controls and sensors which detect liquid level, air and fluid pressures, temperature, and frequency.

Magnetic eddy current drives are used for applications requiring an adjustable speed such as cranes, hoists, fans, compressors, and pumps (figure 13-16).

Variable Pitch Drives

The speed of an ac squirrel cage induction motor depends upon the frequency (hertz) of the supply current and the number of poles of the motor. The equation expressing this relationship is:

$$r/min = \frac{60 \times Hertz}{Pairs\ of\ Poles}$$

A frequency changer may be used to vary the speed of this type of motor. A common method is to drive an alternator through an adjustable mechanical speed drive.

Fig. 13-16 Two magnetic drives driven by 100-hp induction motors mounted on top *(Photo Courtesy Electric Machinery Manufacturing Group.* From Alerich, *Electric Motor Control,* copyright 1983 by Delmar Publishers Inc.)

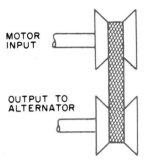

Fig. 13-17 Variable pitch pulley method of obtaining continuously adjustable speed from constant speed shaft (From Alerich, *Electric Motor Control*, copyright 1983 by Delmar Publishers Inc.)

The voltage is regulated automatically during frequency changes. An ac motor drives a variable cone pulley or sheave, which is belted to another variable pulley on the output shaft (figure 13-17). When the relative diameters of the two pulleys are changed, the speed between the input and the output can be controlled. As the alternator speed is varied, the frequency varies, thereby varying the speed of the motor, or motors, connected electrically to the alternator supply. A principal use of this type of drive is on conveyor systems.

Adjustable Frequency AC Drives

Adjustable frequency (static solid-state) drives are also commonly called *inverters*. The power conversion losses are greatly reduced when using these transistor controllers for adjustable speed drives. They are available in a range of horsepowers from fractional to 1,000 hp. Adjustable frequency drives are designed to operate standard ac induction motors. This allows them to be added easily to an existing system.

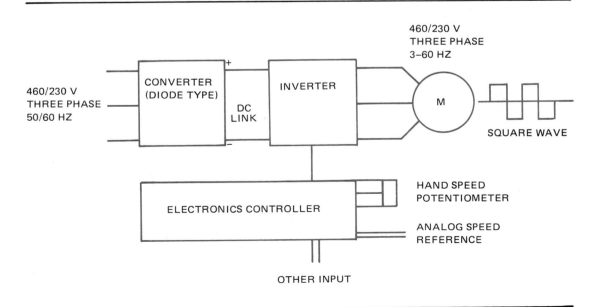

Fig. 13-18 Controller operation

Where energy saving is a major concern the drives are ideal for pumping and fan applications. They are also used for many process control or machine applications where performance is a major concern. Many adjustable speed precision applications were limited to the use of dc motors. By using adjustable frequency controllers with optional dynamic braking, standard squirrel cage motors can now be used in these applications. Municipal, industrial, commercial and mining applications include: sewage, waste water, slurry and booster pumps; ventilation and variable air volume fans; conveyors; production machines and compressors.

Operation. The basic drive consists of the inverter controller which converts the 60-hertz incoming power to a variable frequency and variable voltage. The variable frequency is the actual requirement which will control the motor speed, as shown with the mechanical variable pitch (frequency) drive.

In figure 13-18, the function of the converter is to change the ac input line voltage to a fixed dc voltage. This conversion is accomplished by six power diodes connected in a three-phase, full-wave bridge rectifier configuration (figure 13-19).

The dc link includes capacitors to filter the converter output voltage to a smooth dc voltage. The filter circuit inductors help protect the inverter power switching devices from short circuits or ground faults by limiting the rate at which dc current can fluctuate.

The inverter section changes dc voltage to three-phase ac voltage at the selected frequency. The inverter consists of six gate-turn-off solid-state devices. These devices are switched by the controller electronics to produce a pulse output waveform. The controller electronics includes all circuitry necessary to control output frequency, voltage, and current by generating turn on and turn off signals for the inverter's gate turn-off devices. In addition, light-emitting diodes (LEDs) in the controller electronics monitor and display the controller operating status.

Control and logic functions are integrated on a single, large-scale integration (LSI) chip (figure 13-20). This chip contains approximately 6,300 transistors to perform these complex logic and control functions. These integrated circuits cannot be tested with simple electrical test equipment. Most LSIs must be tested by connecting power to them and then testing the inputs and outputs with special test equipment.

Most industrial controllers are designed with different sections of the control system built in modular form. The electrician determines which section of the circuit

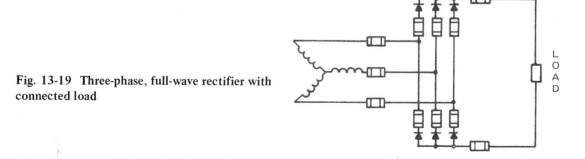

Fig. 13-19 Three-phase, full-wave rectifier with connected load

Fig. 13-20   LSI chip provides control functions in an ac adjustable frequency motor drive   *(Photo courtesy of Allen-Bradley Company, Drives Division)*

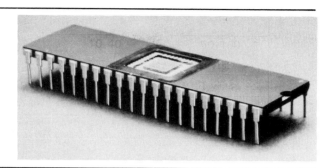

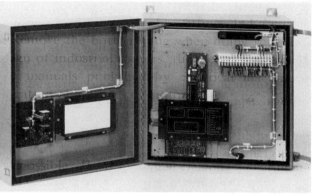

Fig. 13-21   Adjustable frequency motor drive controller   *(Photos courtesy of Allen-Bradley Company, Drives Division)*

is not operating and replaces that module, usually with a spare. The defective module is then sent out to a specialty shop for repair.

Figure 13-21 shows a 1-10 hp adjustable frequency motor drive controller.

ACHIEVEMENT REVIEW

1. What is meant by the term jogging? _____

   _____

   _____

   _____

2. What is meant by the term plugging? _____

   _____

   _____

3. How is dynamic braking applied to an ac induction motor? _____

   _____

   _____

4. How is dynamic braking applied to a synchronous motor? _____

   _____

   _____

5. Draw a schematic diagram of the connections for the control circuit of an across-the-line magnetic motor switch with jogging capability. Include the main relay coil, the pushbutton station with start, jog, and stop pushbuttons, and the sealing contactor.

6. What identifying information should appear on a motor controller so that the controller complies with the requirements of the National Electrical Code?

_____

_____

7. Draw a schematic diagram of an automatic controller used for a three-phase, wound-rotor induction motor.

8. A three-phase, squirrel-cage induction motor has the following ratings: 15 horsepower, 230 volts, 42 amperes per terminal, 40 degrees Celsius, and code classification F. In the spaces in the following table, insert the correct values for fuse protection and running overcurrent protection for this motor when used with each of the types of controllers listed.

| | Fuse Protection | | Running Overcurrent Protection |
Type of Controller	Nontime-Delay	Time-Delay (Dual Element)	
a. Resistance starter			
b. Automatic autotransformer compensator			
c. Across-the-line magnetic motor starting switch with jogging capability			
d. Across-the-line magnetic motor starting switch with plugging capability			

9. What is the purpose of an automatic autotransformer starting compensator?

_____

_____

_____

10. What is the purpose of an automatic controller used with wound-rotor induction motors? _____

_____

_____

11. A wye-delta controller starts the motor at _____
    a. 173 percent of the line voltage.
    b. 58 percent of the line voltage.
    c. full line voltage.
    d. 25 percent of the line voltage.

12. A three-phase, wye-connected motor _____
    a. should never be started by a wye-delta controller.
    b. should always be started by a wye-delta controller.
    c. can be started by a wye-delta controller if proper timers are used.
    d. can be started by a wye-delta controller if proper pushbuttons are used.

13. How are SCRs connected to pass and control ac? _____

_____

_____

14. If a solid-state controller has contactors in the power circuit, in what position are the contacts of start and run in the off position? Why? _____

_____

_____

_____

15. Why are adjustable speed drives used? _____

_____

_____

_____

_____

16. List the types of ac adjustable speed drives. _____

_____

_____

_____

_____

17. How is the speed of the wound rotor motor adjusted? _____

_____

_____

_____

_____

18. How is the eddy current drive coupled to the load? _____

_____

_____

_____

19. How is the ac frequency varied in the mechanical method drive? _____

_____

_____

_____

_____

20. What is the formula for calculating ac motor speed? _____

_____

_____

_____

21. What basic devices are provided in the adjustable frequency drive? _____

_____

_____

_____

_____

22. With an apparent high degree of skill required to maintain an adjustable frequency drive control, how does the plant electrician repair one? _____

_____

_____

_____

_____

THREE-PHASE MOTOR INSTALLATIONS

OBJECTIVES

After studying this unit, the student will be able to

- determine, for several types of three-phase ac induction motors, the
    size of the conductors required for three-phase, three-wire branch circuits.
    sizes of fuses used to provide starting protection.
    disconnecting means required for the motor type.
    size of the thermal overload units required for running overcurrent protection.
    size of the main feeder to a motor installation.
    overcurrent protection required for the main feeder.
    main disconnecting means for the motor installation.
- use the National Electrical Code.

The work of the industrial electrician requires a knowledge of the National Electrical Code requirements which govern three-phase motor installations and the ability to apply these requirements to installations. The elements of a motor circuit are shown in figure 14-1.

This unit gives the procedure for determining the wire size and the proper overload and starting protection for a typical three-phase motor installation. The motor installation example consists of a feeder circuit feeding three branch circuits. Each of these three branch circuits is connected to a three-phase motor of a specified horsepower rating. The feeder circuit and the branch circuits have the necessary overcurrent protection required by the National Electrical Code.

THREE-PHASE MOTOR LOAD

The industrial motor installation described in this example is connected to a 230-volt, three-phase, three-wire service (figure 14-2). The load of this system consists of the following branch circuits.

1. One branch circuit which feeds a three-phase, squirrel-cage induction motor rated at 230 volts, 28 amperes, 10 hp, with a code letter F marking.

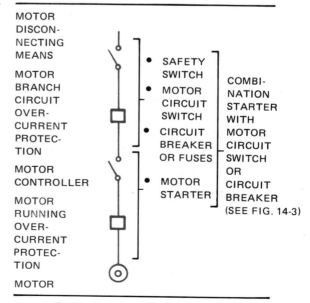

Fig. 14-1 Motor circuit elements

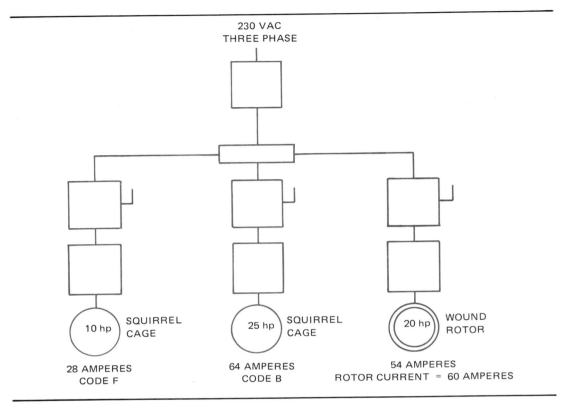

Fig. 14-2 **Branch circuit for each motor**

2. One branch circuit which feeds a three-phase, squirrel-cage induction motor rated at 230 volts, 64 amperes, 25 hp, with a code letter B marking.
3. One branch circuit which feeds a three-phase, wound-rotor induction motor rated at 230 volts, 54 amperes, and 20 hp. The full-load rotor current is 60 amperes.

BRANCH CIRCUIT FOR EACH MOTOR

The values given in NEC *Tables 310-16, 310-17, 310-18* and *310-19,* including notes, shall be used with full-load current for motors in determining ampacity of conductor and fuse size.

Three specific facts must be determined for each of the three branch circuits comprising the load of the installation.

1. The size of the conductors for each three-phase, three-wire branch circuit.
2. The fuse size to be used for starting protection. The fuses protect the wiring and the motor from any faults or short circuits in the wiring or motor windings.
3. The size of the thermal overload units to be used for running protection. The overload units protect the motor from potential damage due to a continued overload on the motor.

NOTE: The full-load amperes shall be taken from the motor's nameplate for calculating thermal overload units (See NEC *Article 430-6*).

BRANCH CIRCUIT 1

The first branch circuit feeds a three-phase, squirrel-cage induction motor. The nameplate data of this motor is as follows:

Squirrel-Cage Induction Motor	
Volts 230	Amperes 28
Phase 3	Speed 1,735 r/min
Code Classification F	Frequency 60 Hertz
10 Horsepower	Temperature Rating 40° Celsius

Conductor Size

Section 430-22(a) of the Code states that branch-circuit conductors supplying a single motor shall have a carrying capacity equal to not less than 125 percent of the full-load current rating of the motor. This general rule must be modified according to *Table 430-22(a) Exception* for certain special service classifications.

The following procedure is used to determine the size of the conductors of the branch circuit feeding the 10-hp motor.

    a. The 10-hp motor has a full-load current rating of 28 amperes. According to *Section 430-22(a)*:

$$28 \times 125\% = 35 \text{ amperes}$$

    b. *Table 310-16* indicates that a No. 8 TW conductor or equivalent is to be used.

    c. Then, *Table 3A, Chapter 9* indicates that three No. 8 TW conductors may be installed in a 3/4-inch conduit.

The squirrel-cage induction motor is to be connected directly across the rated line voltage through an across-the-line motor switch. The branch-circuit overcurrent protection for this motor consists of three standard nontime-delay fuses enclosed in a safety switch located on the line side of the magnetic starter. According to *Section 430-109* of the Code, this switch shall be a motor-circuit switch with a horsepower rating.

NOTE: The Underwriters' Laboratories, Inc. *Electrical Construction Materials List* states that "some enclosed switches have dual horsepower ratings, the larger of which is based on the use of fuses with time delay appropriate for the starting characteristics of the motor. Switches with such horsepower ratings are marked to indicate this limitation and are tested at the larger of the two ratings."

Motor Branch-Circuit Protection

The branch-circuit overcurrent protection for a three-phase, squirrel-cage induction motor marked with the code letter F is given in *Table 430-152*. For the branch circuit 1 motor being considered, the motor circuit overcurrent device shall not exceed 300 percent of the full-load current of the motor (nontime-delay fuses).

The branch-circuit fuse protection for the branch circuit feeding the squirrel-cage motor is:

Since the 10-hp motor has a full-load current rating of 28 amperes, and given the appropriate value from *Table 430-152:*

$$28 \times 300\% = 84 \text{ amperes}$$

Section 430-52 states that if the values for branch-circuit protective devices as determined using the percentages in *Table 430-152* do not correspond to the standard device sizes or ratings, then the next higher size rating or setting may be used.

Section 240-6 of the Code indicates that the next higher standard size fuse above 84 amperes is 90 amperes. Standard nontime-delay cartridge fuses rated at 90 amperes may be used as the branch-circuit protection for this motor circuit.

Disconnecting Means

According to the table for safety switches (figure 14-3) the disconnecting means for this 10-hp motor is a 15-hp, 100-ampere safety switch in which the 90-ampere fuses are installed.

Since these safety switches are dual rated, it is permissible to install a 60-ampere safety switch having a maximum rating of 15 hp if the time-delay fuses are appropriate for the starting characteristics of the motor. The size of the time-delay fuses installed in the safety switch depends on the degree of protection desired and the type of service required of the motor. Time-delay fuses ranging in size from 35 amperes to 60 amperes may be installed in the safety switch (figure 14-4).

Running Overcurrent Protection

The running overcurrent protection consists of three small thermal units housed in the across-the-line magnetic motor starter. (See the note following *Table 430-37* of the Code for an exception to this statement.)

Three-Pole, Three-Fuse, 230-Volt Ac Safety Switches		
	Approximate Manufacturer Horsepower Ratings	
Amperes	Standard	Maximum
30	3	7 1/2 *
60	7 1/2	15 *
100	15	30 *
200	25	60 *
400	50	100 *

*Maximum ratings may be used with time-delay fuses when these fuses are appropriate for the specific motor.

Fig. 14-3 Table for safety switches

Section 430-32(a)(1) of the Code states that the running overcurrent protection for a motor shall be not more than 125 percent of the full-load current (as shown on the nameplate) for motors with a marked temperature rise not over 40 degrees Celsius.

The current rating of the thermal units used as running overcurrent protection is:

$$28 \times 125\% = 35 \text{ amperes}$$

When the selected overload relay is not sufficient to start the motor or to carry the load, *Section 430-34* permits the use of the next higher size or rating, but not higher than 140 percent of the full-load motor current.

BRANCH CIRCUIT 2

A second branch circuit feeds a three-phase, squirrel-cage induction motor. The nameplate data for this motor is as follows:

Fig. 14-4 **Combination fused disconnect switch and motor starter** *(Photo courtesy of Square D Company)*

Squirrel-Cage Induction Motor	
Volts 230	Amperes 68
Phase 3	Speed 1,740 r/min
Code Classification B	Frequency 60 Hertz
25 Horsepower	
Temperature Rating 40° Celsius	

Conductor Size

The following procedure is used to determine the size of the conductors of the branch circuit feeding the 25-horsepower motor.

    a. The 25-hp motor has a full-load current rating of 68 amperes (see NEC *Table 430-150*). According to Code *Section 430-22(a)*:

$$68 \times 125\% = 85 \text{ amperes}$$

    b. *Table 310-16* indicates that a No. 3 Type TW copper conductor, a No. 4 Type THW conductor, or a No. 4 Type RH conductor can be used.

    c. *Table 3A* of *Chapter 9* shows that three No. 3 TW conductors may be installed in a 1 1/4-inch conduit. A 1-inch conduit is required for three No. 4 THW conductors.

NOTE: *Section 373-6(c)* of the Code requires that where conductors of No. 4 size or larger enter an enclosure, an insulating bushing or equivalent must be installed on the conduit.

Motor Branch-Circuit Protection

The 25-hp squirrel-cage induction motor is to be started using a starting compensator. The branch-circuit overcurrent protection for this motor circuit consists of three nontime-delay fuses located in a safety switch mounted on the line side of the starting compensator.

For a squirrel-cage induction motor which is marked with code letter B and which is being used with a starting compensator, *Table 430-152* of the Code requires that the branch-circuit overcurrent protection not exceed 250 percent of the full-load current of the motor.

The branch-circuit overcurrent protection for the branch circuit feeding this motor is:

Since the 25-hp motor has a full-load current rating of 68 amperes (NEC *Table 430-150*),

$$68 \times 250\% = 170 \text{ amperes}$$

Section 240-6 does not show 170 amperes as a standard size for a fuse. However, *Section 430-52* permits the use of a fuse of the next higher size which, in this case, is 175 amperes. Therefore, three 175-ampere nontime-delay fuses are used as the branch-circuit protection for this motor.

Disconnecting Means

According to the table for safety switches in figure 14-3, the disconnecting means for the 25-hp motor is a 25-hp, 200-ampere safety switch in which the 175-ampere fuses are installed.

Because of the dual horsepower rating feature of the safety switches, it is possible to install a 100-ampere safety switch having a maximum rating of 30 hp. The size of the time-delay fuses required in this switch depends on the degree of protection desired and the type of service demanded of the motor. Time-delay fuses ranging in size from 70 amperes to 100 amperes may be installed in the safety switch.

Running Overcurrent Protection

The running overcurrent protection consists of three magnetic overloads located in the starting compensator. According to the nameplate, the motor has a full-load current rating of 68 amperes. The current setting of the magnetic overload units is:

$$68 \times 125\% = 85 \text{ amperes}$$

BRANCH CIRCUIT 3

A third branch circuit feeds a three-phase, wound-rotor induction motor. The nameplate data for this motor is as follows:

Wound-Rotor Induction Motor	
Volts 230	Stator Amperes 54
Phase 3	Rotor Amperes 60
Frequency 60 Hertz	20 Horsepower
Temperature Rating 40° Celsius	

Conductor Size (Stator)

The following procedure is used to determine the size of the conductors of the branch circuit feeding the 20-horsepower motor.

a. The 20-hp motor has a full-load current rating of 54 amperes. According to NEC *Section 430-22(a)*, and *Table 430-150,*

$$54 \times 125\% = 67.5 \text{ amperes}$$

b. *Table 310-16* indicates that a No. 4 Type TW conductor (70 amperes) or a No. 6 Type THHN conductor (70 amperes) can be used.

c. *Tables 3A* and *3B* of *Chapter 9* show that three No. 4 TW conductors may be installed in a 1-inch conduit, or three No. 6 THHN conductors may be installed in a 3/4-inch conduit.

NOTE: *Section 373-6(c)* requires that where conductors of No. 4 size or larger enter an enclosure, an insulating bushing or equivalent must be installed on the conduit.

Motor Branch-Circuit Protection

The 20-hp wound-rotor induction motor is to be started by means of an across-the-line magnetic motor switch. This motor starter applies the rated three-phase voltage to the stator winding. Speed control is provided by a manual drum controller used in the rotor or secondary circuit. All of the resistance of the controller is inserted in the rotor circuit when the motor is started. As a result, the inrush starting current to the motor is less than if the motor were started at the full voltage.

The branch-circuit overcurrent protection of a wound-rotor induction motor is required by *Table 430-152* of the Code not to exceed 150 percent of the full-load running current of the motor.

The branch-circuit overcurrent protection for the branch circuit feeding this motor is:

The full-load current equals 54 amperes for a 20-hp wound-rotor motor

$$54 \times 150\% = 81 \text{ amperes}$$

Section 240-6 does not show 81 amperes as a standard fuse size. Since *Section 430-52* permits the use of a fuse of the next higher size, 90-ampere nontime-delay fuses (*Section 240-6*) can be used.

Disconnecting Means

According to the table for safety switches in figure 14-3, the disconnecting means for the 20-hp motor is a 25-hp, 200-ampere safety switch. Reducers must be installed in this switch to accommodate the 90-ampere fuses required for the motor branch-circuit protection. Because of the dual rating of these safety switches, it is permissible to use a 100-ampere switch having a maximum rating of 30 hp. In this case, standard 90-ampere nontime-delay fuses or 90-ampere time-delay fuses can be installed.

Running Overcurrent Protection

The running overcurrent protection consists of two or three thermal overload units located in the across-the-line magnetic motor starter (except as indicated in the note following *Table 430-37*). According to the nameplate, the motor has a full-load current rating of 54 amperes. The current rating of each thermal unit is:

$$54 \times 125\% = 67.5 \text{ amperes}$$

Conductor Size (Rotor)

The rotor winding of the 20-hp, wound-rotor induction motor is rated at 60 amperes. The following procedure is used to determine the size of the conductors for the secondary circuit from the rotor slip rings to the drum controller.

a. *Section 430-23(a)* requires that the conductors connecting the secondary of a wound-rotor induction motor to its controller have a current-carrying capacity not less than 125 percent of the full-load secondary current of the motor for continuous duty.

$$60 \times 125\% = 75 \text{ amperes}$$

b. *Table 310-16* indicates that several types of copper conductors can be used: No. 3 Type TW, No. 4 Type THW, or No. 6 Type THHN.

c. *Table 3A* of *Chapter 9* shows that three No. 3 TW conductors can be installed in a 1 1/4-inch conduit. A 1-inch conduit is required if three No. 4 THW conductors are used. A 3/4-inch conduit is required for three No. 6 THHN wires.

NOTE: *Section 373-6(c)* requires the use of insulating bushings or equivalent on all conduits containing conductors of No. 4 size or larger entering enclosures.

If the resistors are mounted outside the speed controller, the current capacity of the conductors between the controller and the resistors shall be not less than the values given in *Table 430-23(c)*.

For example, the manual speed controller used with the 20-hp wound-rotor induction motor is to be used for heavy intermittent duty. *Section 430-23(c)* requires that the conductors connecting the resistors to the speed controller have an ampacity not less than 85 percent of rated rotor current.

$$60 \times 85\% = 51 \text{ amperes}$$

Table 310-16 indicates that 51 amperes can be carried safely by No. 8 or No. 6 wire depending on the type of conductor insulation used. As a result, the temperatures generated at the resistor location are an important consideration.

Section 430-32(d) states that the secondary circuits of wound-rotor induction motors, including the conductors, controllers, and resistors, shall be considered as protected against overload by the motor running overcurrent protection in the primary or stator circuits. Therefore, no overcurrent protection is necessary in the rotor or secondary circuit.

MAIN FEEDER

When the conductors of a feeder supply two or more motors, the required wire size is determined using Code rules. *Section 430-24* of the Code states that the feeder shall have an ampacity of not less than 125 percent of the full-load current of the highest rated motor of the group plus the sum of the full-load current ratings of the remaining motors in the group. The full-load current of the motor is taken from NEC *Table 430-150*.

The motor with the largest full-load running current is the 25-hp motor. This motor has a full-load current rating of 68 amperes. The main feeder size, then, in compliance with *Section 430-24*, is:

$$68 \times 125\% = 85 \text{ amperes}$$

Then: 85 + 54 + 28 = 167 amperes.

Table 310-16 indicates that No. 4/0 Type TW or No. 1/0 Type THHN copper conductors can be used.

Tables 3A and *3B* of *Chapter 9* show that three No. 4/0 TW conductors can be installed in 2-inch conduit. Three No. 1/0 THHN conductors can be installed in a 1 1/4-inch conduit. Remember that *Section 373-6(c)* of the Code requires the use of insulating bushings or equivalent on all conduits containing conductors of No. 4 size or larger entering enclosures.

Main Feeder Short-Circuit Protection

Section 430-62(a) states that a feeder which supplies motors shall be provided with overcurrent protection. The feeder overcurrent protection shall not be greater than the largest current rating of the branch-circuit protective device for any motor of the group, based on *Table 430-152*, plus the sum of the full-load currents of the other motors of the group.

The branch circuit feeding the 25-hp motor has the largest value of overcurrent protection. This value, as determined from *Table 430-152*, is 170 amperes (68 × 250% = 170 amperes.) The next standard size fuse is 175 amperes.

The full-load current rating of the 20-hp motor is 54 amperes, and the full-load current rating of the 10-hp motor is 28 amperes. The size of the fuses to be installed in the main feeder circuit is equal to the sum (or next highest value) of 170 + 54 + 28 = 252 amperes.

Therefore, three 250-ampere nontime-delay fuses are used for the feeder circuit. This procedure should be in conformance with *Example 8, Chapter 9* of the Code.

Main Disconnecting Means

Section 430-109 lists several exceptions to the ruling that the disconnecting means shall be a motor-circuit switch, rated in horsepower, or a circuit breaker. The disconnecting means shall have a carrying capacity of at least 115 percent of the nameplate current rating of the motor, *Section 430-110(a)*. Therefore, the 250-ampere fuses required as the overcurrent protection for the main feeder are installed in a 400-ampere safety switch.

Wire types and sizes are selected by the ambient temperatures of the place of installation and the economics of the total installation, such as the minimum size conduits, cost of the wire sizes, and the cost of the labor to install the different selections.

ACHIEVEMENT REVIEW

A feeder circuit feeds three branch motor circuits. Branch motor circuit No. 1 has a load consisting of an induction motor with the following nameplate data:

No. 1

Squirrel-Cage Induction Motor	
230 Volts	15 Amperes
3 Phase	60 Hertz
5 Horsepower	Code Classification D
Temperature Rating 40° Celsius	

Branch motor circuit No. 2 has a load consisting of an induction motor with the following nameplate data: (This motor is equipped with an autotransformer starting compensator.)

No. 2

Squirrel-Cage Induction Motor	
230 Volts	40 Amperes
3 Phase	60 Hertz
15 Horsepower	Code Classification F
Temperature Rating 40° Celsius	

Branch motor circuit No. 3 has a load consisting of a wound-rotor induction motor with the following nameplate data:

No. 3

Wound-Rotor Induction Motor	
230 Volts	22 Stator Amperes
3 Phase	26 Rotor Amperes
7 1/2 Horsepower	60 Hertz
Continuous Duty	
Temperature Rating 40° Celsius	

1. Refer to the following diagram.

   a. Determine the running overload protection in amperes required for the motor in branch circuit No. 1.

b. Determine the appropriate wire size (TW).

(Insert the answers on the diagram.)

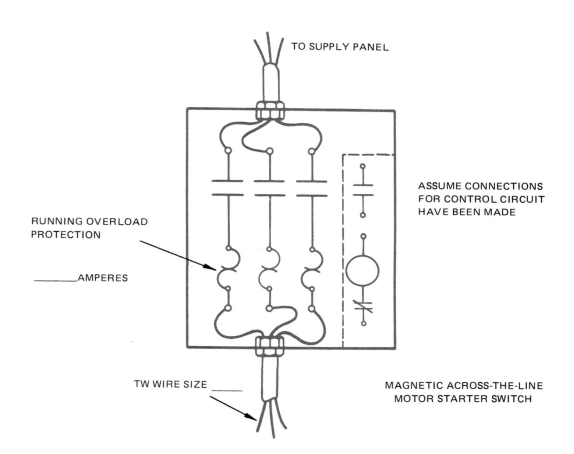

TO SUPPLY PANEL

ASSUME CONNECTIONS
FOR CONTROL CIRCUIT
HAVE BEEN MADE

RUNNING OVERLOAD
PROTECTION

_____AMPERES

TW WIRE SIZE _____

MAGNETIC ACROSS-THE-LINE
MOTOR STARTER SWITCH

2. Refer to the following diagram.

   a. Determine the running overload protection in amperes required for the motor in branch circuit No. 2.

   b. Determine the appropriate wire size of the copper TW conductors. Note that the 15-hp squirrel-cage induction motor in this circuit is started by means of a starting compensator.

   (Insert the answers on the diagram.)

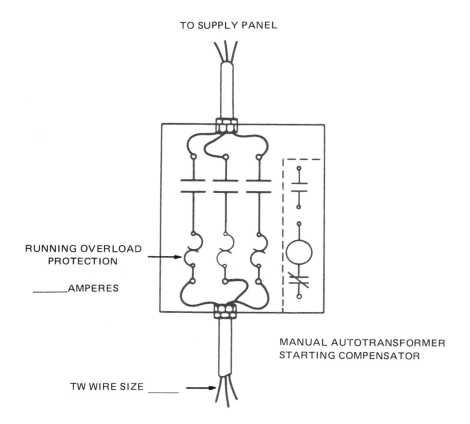

TO SUPPLY PANEL

RUNNING OVERLOAD
PROTECTION

_____AMPERES

MANUAL AUTOTRANSFORMER
STARTING COMPENSATOR

TW WIRE SIZE _____

3. Refer to the following diagram.

  a. Determine the running overload protection in amperes required for the motor in branch circuit No. 3.

  b. Determine the appropriate wire size of the copper conductors.

  (Insert the answers on the diagram.)

  c. Determine the size of the conductors required for the secondary circuit of the wound-rotor induction motor in branch circuit No. 3. The secondary or rotor circuit feeds between the slip rings of the wound rotor and the speed controller. Indicate the size of the conduit. Use TW conductors. _____

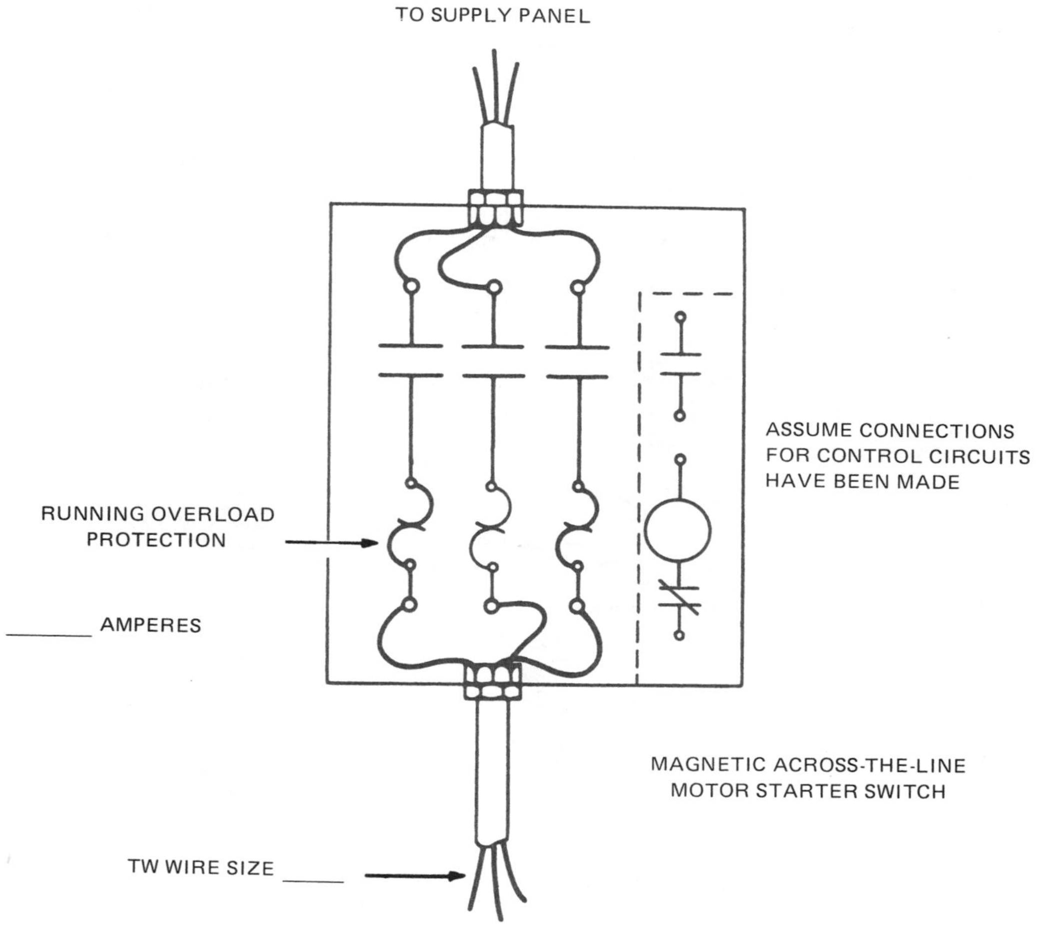

TO SUPPLY PANEL

ASSUME CONNECTIONS
FOR CONTROL CIRCUITS
HAVE BEEN MADE

RUNNING OVERLOAD
PROTECTION

_____ AMPERES

MAGNETIC ACROSS-THE-LINE
MOTOR STARTER SWITCH

TW WIRE SIZE _____

4. Refer to the following diagram.

   a. Determine the current rating in amperes of the fuses (nontime-delay) used as overload protection for the main feeder circuit shown in the diagram.

   b. Determine the TW conductor size for the main feeder switch.

   (Insert the answers on the diagram.)

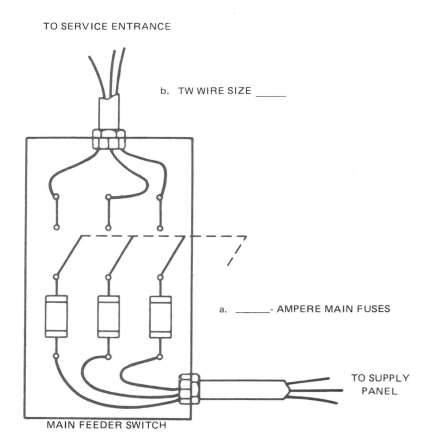

TO SERVICE ENTRANCE

b. TW WIRE SIZE _____

a. _____- AMPERE MAIN FUSES

TO SUPPLY PANEL

MAIN FEEDER SWITCH

5. Refer to the following diagram.

a. Using TW-type copper conductors, determine the size of the conductors and conduit required for the main feeder circuit which feeds the three branch motor circuits. Indicate the sizes on the diagram.

b. Determine the size of fuses in amperes required for the starting overload protection for each of the branch circuits.

Motor Circuit No. 1 _____

Motor Circuit No. 2 _____

Motor Circuit No. 3 _____

(Insert the answers on the diagram.)

c. Using TW-type copper conductors, determine the size of rigid conduit required for each of the three branch circuits.

Motor Circuit No. 1 _____

Motor Circuit No. 2 _____

Motor Circuit No. 3 _____

(Insert the answers on the diagram.)

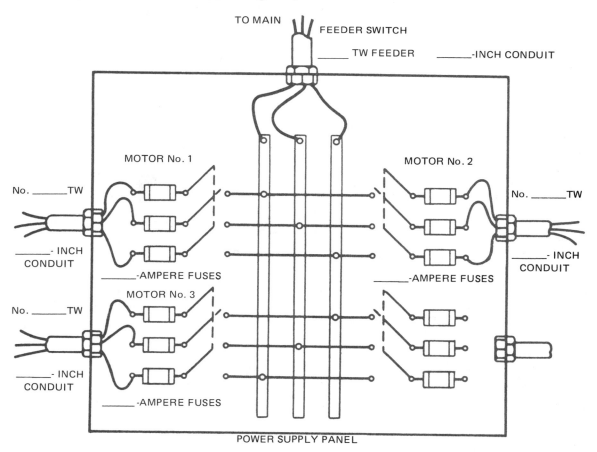

MOTOR MAINTENANCE

OBJECTIVES

After studying this unit, the student will be able to

- perform routine inspection and maintenance checks of three-phase motors.
- perform the following simple tests:
  measure insulation resistance
  use a growler to locate short-circuited coils
  continuity checks for open-circuited coil
  balance test to determine phase currents under load
  speed variation
- replace and lubricate sleeve and ball bearings, according to manufacturers' directions.
- lubricate motors, according to manufacturers' directions.

PREVENTIVE MAINTENANCE

Most electrical equipment requires planned inspection and maintenance to keep it in proper working condition. Periodic inspections prevent serious damage to machinery by locating potential trouble areas. Observant personnel will make full use of their senses to diagnose and locate problems in electrical machinery: the sense of *smell* directs attention to burning insulation; the sense of *feel* detects excessive heating in windings or bearings; the sense of *hearing* detects excessive speed or vibration; the sense of *sight* detects excessive sparking and many mechanical faults.

Sensory impressions usually must be supplemented by various testing procedures to localize the trouble. A thorough understanding of electrical principles and the efficient use of test equipment is important to the electrician in this phase of troubleshooting.

PERIODIC INSPECTIONS

The ideal motor maintenance program aims at preventing breakdowns rather than repairing them. Systematic and periodic inspections of motors are necessary to ensure best operating results. In a good preventive maintenance program with detailed checks, the person in charge should have a record card on file for every motor in the plant. Entries on the card should include inspection dates, descriptions of repairs, and the costs involved. When the record indicates that a motor has undergone excessive and/or costly repairs, the causes can be determined and corrected.

Inspection records also serve as a guide to indicate when motors should be replaced because of their high cost of operation. They also reveal faulty operating conditions, such as misapplication or poor drive engineering.

Inspection and servicing should be systematic. However, the frequency of inspections and the degree of thoroughness may vary, as determined by the plant maintenance engineer. Such determinations are based on 1) the importance of the motor in the production scheme (if the motor fails, will production be slowed seriously, or stopped?), 2) the percentage of the day the motor operates, 3) the nature of the service, and 4) the motor's environment. An inspection schedule, therefore, must be flexible, and adapted to the needs of each plant. Equipment manufacturers' specifications and procedures should be consulted and followed.

The following schedule, which covers both ac and dc motors, is based on average conditions insofar as operational use and cleanliness are concerned. (Where dust and dirty conditions are extremely severe, open motors may require a certain amount of cleaning every day.)

EVERY WEEK

1. Examine commutator and brushes, ac and dc.
2. Check oil level in bearings.
3. See that oil rings turn with shaft.
4. See that exposed shaft is free of oil and grease from bearings.
5. Examine the starter switch, fuses, and other controls; tighten loose connections.
6. See that the motor is brought up to speed in normal time.

EVERY SIX MONTHS

1. Clean motor thoroughly, blowing out dirt from windings, and wipe commutator and brushes.
2. Inspect commutator clamping ring.
3. Check brushes and replace any that are more than half worn.
4. Examine brush holders, and clean them if dirty. Make certain that brushes ride free in the holders.
5. Check brush pressure.
6. Check brush position.
7. Drain, wash out, and replace oil in sleeve bearings.
8. Check grease in ball or roller bearings.
9. Check operating speed or speeds.
10. See that end play of shaft is normal.
11. Inspect and tighten connections on motor and control.
12. Check current input and compare it with normal.
13. Examine drive, critically, for smooth running, absence of vibration, and worn gears, chains, or belts.
14. Check motor foot bolts, end-shield bolts, pulley, coupling, gear and journal set-screws, and keys.
15. See that all covers, and belt and gear guards are in place, in good order, and securely fastened.

ONCE A YEAR

1. Clean out and renew grease in ball or roller bearing housings.

2. Test insulation by megohmmeter.
3. Check air gap.
4. Clean out magnetic dirt that may be clinging to poles.
5. Check clearance between shaft and journal boxes of sleeve bearing motors to prevent operation with worn bearings.
6. Clean out undercut slots in commutator. Check the commutator for smoothness.
7. Examine connections of commutator and armature coils.
8. Inspect armature bands.

MEASUREMENT OF INSULATION RESISTANCE

The condition of insulation is an important factor in the maintenance of motors and alternators. Moisture, dirt, chemical fumes, and iron particles all cause deterioration of the insulation used on the windings of stators and rotors. Motors operating under adverse conditions require periodic tests to insure continuous and satisfactory operation. Although severe conditions can be detected by touch, sight, or smell, often it is necessary to use more accurate measures of the condition of insulation at any given time.

The value of insulation resistance in *megohms* (1,000,000 ohms) is used as an indication of insulation efficiency. Successive readings taken and recorded under the same test conditions will document the insulation history of a unit and will serve as an index of insulation deterioration. A *megohmmeter,* commonly called a Megger® (figure 15-1A and B) is used to measure insulation resistance. In general, the instrument develops a voltage which is applied to the insulation path. The amount of current in this path is shown on a sensitive *microammeter* calibrated in megohms. When the instrument leads are not connected, the microammeter reading should be infinity. Specific operating instructions are provided with the instrument by the manufacturer.

The megohmmeter ground lead is connected to the frame of the machine. The ungrounded lead is connected to any metallic part of the winding being tested, such as any terminal of the coil circuit. No other external paths must exist in parallel circuits. For this reason, the windings being tested should be isolated by disconnecting them from other parts of the circuit. The megohmmeter reading then is the resistance of the insulation between the number of coils in the circuit and the frame of the machine.

Major motor manufacturers apply insulation resistance for a pressure of 660 volts for low-voltage motors below 600 volts. A method of determining the normal insulation resistance value is given in the following formula. It is a nonexact formula derived from experience. It is believed that the Institute of Electrical and Electronic Engineers, Inc. (IEEE) accepted this general theory from a manufacturer of megohmmeters. *Caution:* Disconnect electronic devices attached to units under test.

$$\text{Megohms} = \frac{\text{Rated Voltage of Machine}}{\dfrac{\text{Rating in kVA}}{100} + 1,000}$$

A very low value of insulation resistance indicates defective insulation. The electrician should begin an immediate check to localize the defective insulation. This is done by disconnecting various coils from a series, parallel, wye, or delta combination and re-

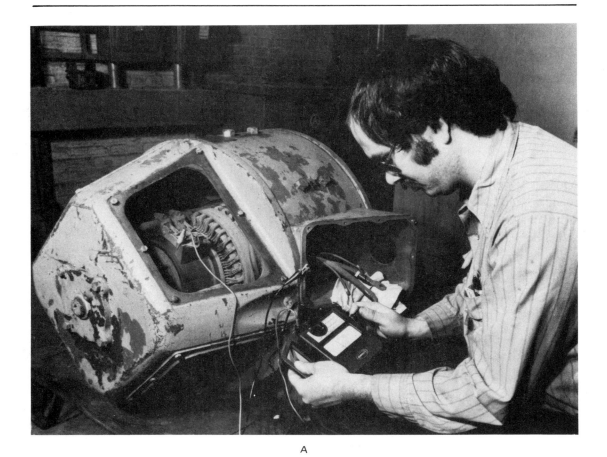

A

B

Fig. 15-1   A) A Megger® tester being used to check the insulation resistance of an electric motor prior to repair   B) Major Megger® insulation tester *(Photos courtesy of Biddle Instruments)*

peating the insulation resistance test on each isolated coil. A very low value of resistance may indicate a grounded coil (complete breakdown of insulation at some point).

A value of insulation resistance slightly below the recommended approximate value does not necessarily indicate that immediate repair is required. The electrician should take a series of readings at weekly intervals to detect any progressive decrease or sudden drop in the insulation resistance. If the resistance continues to decrease, then the fault is to be located without delay. A slightly low but constant value of resistance should not cause concern. NOTE: If a motor has been reading 10 megohms on periodic tests and suddenly drops to 0.2 megohms, the motor may be unsafe to start. New motors read around infinity (the highest resistance on the meter scale), or slightly lower, but, during use, this resistance lowers to a more steady reading.

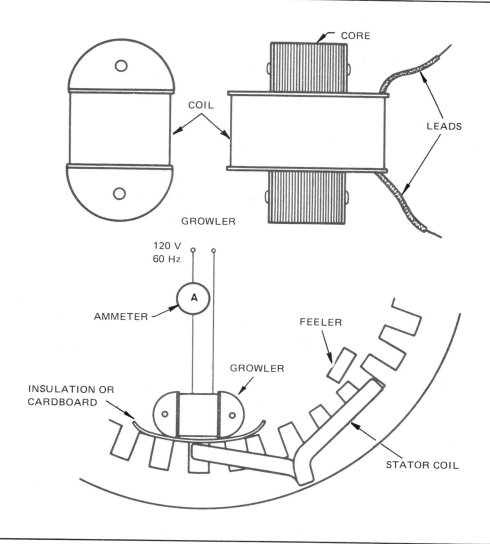

Fig. 15-2 **A growler for testing shorted coils**

TESTING FOR SHORT-CIRCUITED COILS

Open-circuited coils on rotors or stators can be located by continuity (one end to the other) tests. Short-circuited coils are located easily by the use of a growler. A *growler* is an instrument consisting of an electromagnetic yoke and winding excited from an ac source. The yoke is placed across a section of slots containing the winding being tested. The yoke winding acts as the primary winding of a transformer and the winding being tested acts as the secondary (figure 15-2).

If a turn or coil is short circuited, the resulting current rise in the primary (yoke) circuit is indicated on the ammeter. If the current is permitted to exist for a short period of time, the defective turn or coil can be identified by the heat developed at the defective point. The stator windings of both alternators and motors can be tested by this method.

The field coils of alternators can be tested using the voltage drop method described for dc machinery testing. With a given value of current in the field circuit, the voltage drops on individual field coils should be approximately equal. If there is a voltage drop difference in excess of 5 percent, the coil should be investigated for shorted turns. The presence of full line voltage across a single coil indicates an open circuit in that coil.

The field coils of alternators can be checked for impedance by applying a high frequency voltage to each coil and measuring the current. The currents in the coils should be equal. The presence of a high current usually means that there are shorted turns somewhere in the coil.

BALANCE TEST

The current in the individual phases of three-phase motors must be equal. Figure 15-3 shows the connections necessary to make a simple balance test which measures the phase currents under load. This test may be made in the electric shop before a motor is installed.

The fused three-pole switch at the left of the diagram is used to start the motor. The second three-pole switch removes the ammeter from the circuit during the starting

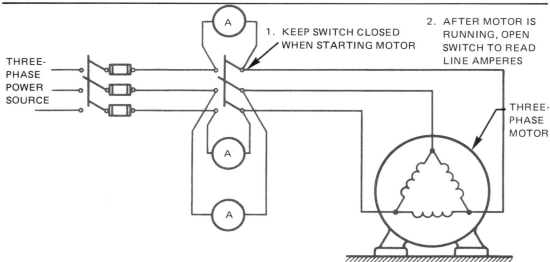

Fig. 15-3 A balanced current motor test

Fig. 15-4 A) Use of a clamp-on ammeter for motor testing *(Photo courtesy of A & M Instruments)*
B) Digital ac clamp-on ammeter *(Photo courtesy of Amprobe Instruments)*

period when the current input is very high. This switch is closed before the motor is
started. When the motor reaches its rated speed, the second switch is opened. The current in each phase, therefore, is indicated on the ammeter. For a motor operating normally, the three line currents are equal. A high reading in one phase may indicate shorted
turns. If one phase shows no current, then the motor is operating as a single-phase
motor. High but equal current readings in all three phases indicate an overloaded motor.

Figure 15-4 shows the use of a more convenient method of making a balance test.
A clamp-on ammeter is used to take readings in each phase of a motor in actual operation under normal load.

MEASUREMENT OF SPEED VARIATIONS

Deviations from the rated speed of a motor under load are an indication of im-
proper mechanical loading or faulty conditions within the motor. A *tachometer* (figures 15-5 and 15-6) is an instrument used to check the speed of a motor. Other types
of instruments are also used. The tachometer shown in figure 15-5 contains tuned carbon steel reeds which vibrate in response to the revolutions per minute (r/min) of a
machine under test. It can register to within 25 r/min. The tachometer in figure 15-6A
is held by hand to the end of the motor shaft, as shown in figure 15-6B.

SQUIRREL CAGE ROTORS

The bars of a squirrel-cage rotor can be broken if the motor is subjected to severe
jarring, vibration, or overheating. Bars that are dislodged must be put back in place
securely so that movement is impossible. The presence of broken bars can be detected
by the use of a growler. (The rotor must be removed to make this test.)

A test can be made which does not require the removal of the rotor to detect broken
or open rotor bars. This test consists of exciting one phase of the stator winding with

Fig. 15-5　Using a vibration response (resonant reed) tachometer.　Note the reading of 1,750 revolutions per minute　*(Photo courtesy of Biddle Instruments)*

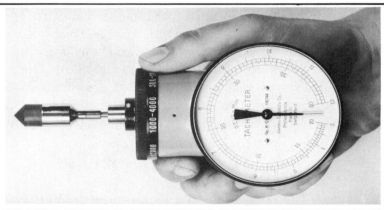

Fig. 15-6　A) A hand tachometer　B) Measuring speed with a hand tachometer　*(Photos courtesy of Biddle Instruments)*

25 percent of its normal voltage. Enough voltage is used to give a suitable indication on an ammeter in series with the winding. By turning the rotor slowly by hand, any variations in stator current can be observed on the ammeter. Any current variation in excess of 3 percent usually indicates open bars in the rotor.

BEARINGS

The type of bearings used in a motor depends on the cost of the bearing and the characteristics of the load. Sleeve and ball bearings are used in both ac and dc motors. Excessive wear on the bearings reduces the concentricity of the stator and rotor sections. In small motors, with the power off, excessive wear can be detected by manually attempting to move the shaft of the rotating member in a lateral direction. The amount of play in the shaft is an indication of bearing wear. For large motors, bearing wear and the resulting deviation in concentricity of the stator and rotor can be detected by measuring the air gaps over several points around the periphery of the gap. Severe bearing wear on both large and small motors may result in actual contact between the rotor and the stator.

The motor must be disassembled to repair bearings. This type of repair requires special tools and the job should not be attempted without them.

A sleeve bearing is removed by dismantling the baffles inside the end shield, removing the oil well cover plates, and removing the oil ring clips. The bearing lining is then tapped out using a short length of pipe stock or a special split fitting which locks inside the oil ring slot.

Ball Bearings

Ball bearings are press fitted to the motor shaft and usually should be removed from the shaft only when it is necessary to replace the bearing.

To inspect the ball bearings, the end bells are removed and the rotor, the rotor shaft, and the bearing assembly are taken from the stator. In some motors, the bearing housings have removable bearing caps so that it is possible to remove the bearing without removing the end bells.

If a ball bearing must be replaced, a bearing puller usually is used to prevent damage to the shaft. The electrician must be very careful when placing a new bearing on the shaft so that neither the bearing nor the shaft is damaged (figure 15-7). A ball bearing race must be placed on the shaft so that the race is exactly square with the shaft. The shaft must be in perfect condition since even the slightest burr will cause trouble. **Pressure must not be applied to the outer race of the bearing.** Pressure applied to install the bearing must be applied evenly on the diameter of the inner race (figure 15-8). Light tapping is recommended. A piece of pipe stock slightly greater than the shaft diameter is used to press on the new bearing. The bearing can be warmed to a temperature of 150°F to simplify the process.

New ball bearings must not be cleaned prior to installation. Dust or dirt must not enter the bearing during the installation.

WRONG:

BEARING SHOULD NOT
BE FORCED ON SHAFT
BY TAPPING ON THE
OUTER RINGS. IT
SHOULD NOT BE FORCED
ON A BADLY WORN
SHAFT OR ON A SHAFT
THAT IS TOO LARGE.

RIGHT:

BEARING IS PROPER
SIZE FOR SHAFT AND
IS BEING TAPPED
LIGHTLY INTO PLACE
BY MEANS OF A METAL
TUBE THAT FITS
AGAINST THE INNER
RING. DO NOT POUND
ON THE BEARING.

Fig. 15-7 Bearing installation

LUBRICATION

Several methods of lubricating motors are used. Small motors with sleeve bearings have oil holes with spring covers. These motors should be oiled periodically with a good grade of mineral oil. Oil with a viscosity of 200 seconds Saybolt (approximately SAE 10) is recommended.

The bearings of larger motors often are provided with an oil ring which fits loosely in a slot in the bearing. The oil ring picks up oil from a reservoir located directly under the ring. Under normal operating conditions, the oil should be replaced in the motor at least once a year. More frequent oil replacement may be necessary in motors operating under adverse conditions. In all cases, avoid excessive oiling; insufficient oil can ruin a bearing but excessive oil can cause deterioration of the insulation of a winding.

Fig. 15-8   Single-row, snap ring ball bearing *(Photo courtesy of General Electric Company)*

Many motors are lubricated with grease. Periodic replacement of the grease is recommended. In general, the grease should be replaced whenever a general overhaul is indicated, or sooner if the motor is operated under severe operating conditions.

Grease can be removed by using a light mineral oil heated to 165°F or a solvent. Any grease-removing solvents should be used in a well-ventilated work area.

Ball Bearing Lubrication

The following table indicates recommended intervals between the regreasing of ball bearing units. It is important that the correct amount and type of grease be used in ball bearings. Too much grease can cause overheating.

Horsepower			
Service	1/4-7 1/2	10-40	50-150
Easy	7 yrs.	5 yrs.	3 yrs.
Standard	5 yrs.	2 yrs.	1 yr.
Severe	3 yrs.	1 yr.	6 mos.
Very severe	6 mos.	3 mos.	3 mos.

Fractional horsepower motors often contain sealed bearings. With normal service, these bearings do not require regreasing. When regreasing is required for unshielded bearings, the manufacturers' specifications and directions must be followed.

Some motors with ball bearings are provided with pressure fittings and a grease gun is used to lubricate the bearings. Remove the bottom plug when doing this. Because of the wide variation in the design of industrial motors, the electrician should consult the comprehensive lubrication manuals published by electrical machinery manufacturers to insure the proper lubrication of all types of motors.

BALL AND ROLLER BEARING MAINTENANCE

Cleaning Out Old Grease

It is best to remove the bearing, if possible. However, when cleaning a bearing in place, remove as much of the old grease as possible, using a rag or brush free from dirt or dust. Flush out the housing with clean, hot kerosene (110°F to 125°F); clean, new oil; or solvent.

After the grease has been removed, flush out the bearing with light mineral oil to prevent rust and to remove all traces of cleaning fluid. Allow to drain thoroughly before adding new grease.

When the bearing is removed from the case, wash out the hardened, rancid grease from the housing and bearing as follows:

1. Put on safety glasses, or a face shield, for protection.
2. Soak the bearing in hot kerosene, then remove it.
3. Rotate the bearing slowly by air hose. (Fast rotation may score the balls without lubrication.)
4. Dip the bearing in clean kerosene, light oil, or solvent.
5. Rotate the bearing slowly again by air hose.
6. Wash the bearing again with clean kerosene or solvent.
7. Rotate the bearing in hand, and check for smoothness.
8. If the bearing is smooth, repack it with grease.
9. If the bearing is not smooth, but is in good condition, there is still some hardened grease in it. Repeat operations 6 and 7.

Lubricating Bearings with Pressure Relief Plug

1. Wipe fitting and plug clean.
2. Remove relief plug in bottom of bearing (figure 15-9).

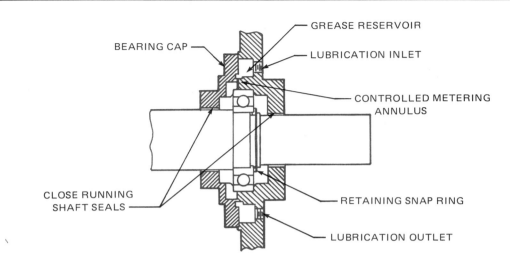

Fig. 15-9 Ball bearing housing assembly. Note the lubrication in-
let and outlet *(Photo courtesy of General Electric Company)*

Fig. 15-10 Gear motor *(Photo courtesy of Sterling Electric, Inc.)*

3. With shaft in motion (if possible), force grease into the top grease fitting, catching old grease in a pan. Add grease until new grease appears at the pan.
4. With a screwdriver, open the relief hole. Do *not* push the screwdriver into the gearing housing.
5. Allow bearing to run with relief plug out to remove pressure.
6. After the grease stops running out, and there is no longer any pressure in the bearing, replace the relief plug.

Lubricating Bearings without Relief Plug

1. Install a grease fitting with safety vents.
2. With the motor running, pump in grease slowly until a slight bleed shows around the seal or safety vent.
3. If necessary to lubricate while bearing is standing, fill the bearing with grease from one-fourth to one-half of its capacity.
   NOTE: It is important not to overgrease, and grease must be kept clean.

Relubrication with Oil

Oil is always subject to gradual deterioration from use, and contamination from dirt and moisture. Because of this, regular intervals for cleaning bearings must be maintained.

1. After draining the used oil, flush out the bearing. This can be done by using a new charge of the lubricant used for regular lubrication. Run the machine for three to five minutes, and drain again. Units used where there is sawdust or dirty conditions may require two flushings.
2. Fill with new oil to the proper level. Be sure new oil is from a clean container, and that no dirt is pushed into the filler plug while filling. Lubricant must be kept clean.
3. Check the seals to see that they are effective to prevent leakage or the entry of outside dirt to the bearing. It is important to keep the lubricant clean and the bearing flushed out so that it will be clean. Always read and follow the manufacturer's instructions whenever possible.

OILS FOR GEAR MOTORS

A gear motor is a self-contained drive made up of a ball bearing motor and a speed reducing gear box (figure 15-10). It is designed to take advantage of the electrical efficiency of the high speed motor and the transmission efficiency of gears.

The front motor bearings are generally grease lubricated and require the same attention as standard ball bearing motors. The rear bearings, gear box bearings, and the gears themselves are almost always splash lubricated from the same oil supply reservoir in the lower section of the gear unit.

Oil seals at each bearing prevent oil leakage into the motor windings and out along the drive shaft. The precision cut gears require carefully selected lubricating oils. Use only top grade oils of the viscosity called for by the manufacturer of the gear motor.

ACHIEVEMENT REVIEW

A. Select the correct answer for each of the following statements and place the corresponding letter in the space provided.

1. Periodic inspection of motors, controls, and other electrical equipment is important because it _____
   a. gives advance notice of impending trouble.
   b. is required by the job standards.
   c. is a requirement of supervision.
   d. completes a day's work.

2. Careful motor troubleshooting requires the use of _____
   a. the sense of smell.   c. hearing and vision.
   b. the sense of feel.   d. all of these.

3. The most accurate method of testing insulation resistance uses a(an) _____
   a. megohmmeter.   c. ohmmeter.
   b. growler.   d. tachometer.

4. Insulation resistance is measured in _____
   a. megawatts.   c. kilohms.
   b. megohms.   d. kilovolts.

5. A very low value of insulation resistance indicates _____
   a. a good operating condition.
   b. a fair operating condition.
   c. an immediate investigation.
   d. that the measuring instrument is at the wrong setting.

6. Short-circuited coils are located efficiently by the use of a(an) _____
   a. megohmmeter.   c. ohmmeter.
   b. growler.   d. clamp-on ammeter.

7. On a balance test for phase currents where one phase shows a higher reading, the probable cause is _____
   a. open turns.
   b. shorted turns.
   c. worn bearings.
   d. the need for rotor balancing.

8. If a three-phase, squirrel-cage motor is operating with current in only two line leads, then _____
   a. the motor is operating as a two-phase motor.
   b. the insulation is overloaded.
   c. there is an open circuit in the stator.
   d. there is an open circuit in the rotor.

9. A properly-sized ball bearing race is mounted on a shaft correctly by tapping
   a. a metal tube on the outer ring.
   b. the inner ring with a hammer.
   c. the outer ring with a hammer.
   d. a metal tube on the inner ring.   _____

10. A ball-bearing, ten-hp, three-phase, 230-volt induction motor operating under very severe conditions should be greased about every
   a. three months.   c. twelve months.   _____
   b. six months.   d. month.

B. Insert the word or phrase to complete each of the following statements.

11. The instrument used to measure insulation resistance is operated with the test leads unconnected. The meter reading should be _____ .

12. The measurement of individual phase currents in the operation of a three-phase induction motor is called a(an) _____ .

13. The rotor and stator concentricity in three-phase motors and alternators can be determined by measurements of the _____ .

14. Ball bearings should be removed from a motor shaft using a bearing _____
    _____ .

15. Pressure should never be applied to the _____ race of a ball bearing.

16. Ball bearings are lubricated with _____ .

17. Insufficient oil can ruin a bearing, but excessive oiling can ruin the _____
    _____ .

SUMMARY REVIEW OF UNITS 6-15

OBJECTIVE

- To give the student an opportunity to evaluate the knowledge and understanding acquired in the study of the previous ten units.

1. List three types of three-phase ac motors.
    a. _____
    b. _____
    c. _____

2. Insert the word or phrase to complete each of the following statements.
    a. The speed of a three-phase induction motor falls slightly from a no-load condition to a full load. This is true of a three-phase induction motor with a _____ _____ rotor.
    b. A speed controller is used only with a three-phase induction motor of the ____ _____ type.
    c. When all of the resistance of the speed controller is inserted in the secondary circuit of a three-phase _____ _____ induction motor, the starting torque is very good.
    d. A three-phase _____ motor is operated with an overexcited dc field to obtain a leading power factor.
    e. The speed of a three-phase _____ motor remains constant from a no-load condition to full load if the operating frequency remains constant.

3. State two advantages of using a squirrel-cage induction motor. _____
   _____
   _____

4. State one disadvantage of using a squirrel-cage induction motor. _____
   _____

5. Explain how the direction of rotation of a three-phase, squirrel-cage induction motor is reversed. _____
   _____
   _____
   _____
   _____

6. A two-pole, 60-Hz, three-phase, squirrel-cage induction motor has a full-load speed of 3,475 r/min. Determine the synchronous speed of this motor. _____

_____

_____

7. Determine the percent slip of the motor in question 6. _____

_____

_____

8. What the the purpose of starting protection for a three-phase motor? _____

_____

_____

_____

9. What is the purpose of running protection for a three-phase motor? _____

_____

_____

10. Show the connection diagram for the nine terminal leads of a wye-connected, three-phase motor rated at 230/460 volts for three-phase operation on 230 volts.

11. Explain how the running overload protection for a three-phase motor rated at more than 1 horsepower would be selected. _____

_____

_____

_____

12. Insert the correct code letter in the following statements.

a. The National Electrical Code requires squirrel-cage induction motors with code marking _____ to have starting protection rated at not over 150 percent of full-load current for a nontime-delay fuse.

b. The National Electrical Code requires squirrel-cage induction motors with code letter markings _____ to have starting protection rated at not over 250 percent of full-load current for a nontime-delay fuse.

c. The National Electrical Code requires squirrel-cage induction motors with _____ code letter markings to have starting protection rated at not over 300 percent of full-load current for a nontime-delay fuse.

d. The National Electrical Code requires squirrel-cage induction motors with auto-transformer starting and code letter markings _____ to have starting protection rated at not over 200 percent of full-load current for a nontime-delay fuse.

13. Why are starting compensators used with large three-phase, squirrel-cage induction motors? _____

_____

14. Why is a wound-rotor induction motor used in place of a squirrel-cage induction motor for some industrial applications? _____

_____

_____

_____

_____

15. Explain how the direction of rotation of a three-phase, wound-rotor induction motor can be reversed. _____

_____

_____

_____

16. Insert the correct word or phrase to complete each of the following statements.

   a. The speed of a wound-rotor induction motor is _____ by inserting resistance in the rotor circuit through a speed controller.

   b. The starting surge of current of a wound-rotor induction motor is minimized by _____ .

   c. The rotation of a wound-rotor induction motor is _____ by changing any two of the three leads feeding from the rotor slip rings to the speed controller.

   d. The _____ of a wound-rotor induction motor is very good if all of the resistance of the speed controller is inserted in the rotor circuit.

   e. The efficiency of a wound-rotor induction motor operating at rated load with all of the resistance inserted in the rotor circuit is _____ .

17. Draw a schematic connection diagram of a wound-rotor induction motor which is started by means of an across-the-line magnetic motor starter controlled from a pushbutton station. Include a wye-connected speed controller in the rotor circuit.

18. Explain how the direction of rotation of a three-phase synchronous motor is reversed.

_____

_____

19. List two important applications for three-phase synchronous motors. _____

_____

_____

20. A three-phase synchronous motor with four stator poles and four rotor poles is operated from a three-phase, 60-Hz line of the correct voltage rating. Determine the speed of the motor. _____

_____

_____

21. Explain the correct procedure for starting a three-phase synchronous motor.

_____

_____

_____

22. Insert the correct word or phrase to complete each of the following statements.

a. The speed of a synchronous motor is _____ from no load to full load.

b. A synchronous motor with an underexcited dc field has a _____ power factor.

c. A three-phase _____ motor must be started as an induction motor.

23. What is an amortisseur winding? _____

_____

_____

24. Explain what is meant by the term jogging. _____

_____

_____

25. Explain what is meant by the term plugging. _____

_____

_____

26. What identifying information should appear on a motor controller to comply with the requirements of the National Electrical Code? _____

_____

_____

_____

_____

27. A three-phase, squirrel-cage induction motor is rated at 25 hp, 230 volts, 64 amperes per terminal, 40 degrees Celsius, and is classified as code letter F. In the following table, fill in the correct fuse size required for branch-circuit protection and the correct running overload protection for this motor, when used with each of the types of controllers listed.

| | Fuse Protection | | Running Overcurrent Protection |
Type of Controller	Nontime-delay	Time-delay	
a. Resistance starter			
b. Automatic autotransformer compensator			
c. Across-the-line magnetic motor starting switch with jogging capability			
d. Across-the-line magnetic motor starting switch with plugging capability			

28. What size of wire and conduit are used for the branch circuit feeding the motor in question 28? (Use Type TW.) _____

_____

29. Insert the correct word or phrase to complete each of the following statements.

a. A controller with _____ may be used to stop a motor quickly.

b. When fuses are used as protection for a three-phase, three-wire ungrounded branch motor circuit, the fuses must be installed in _____ line leads.

c. Motors operating on a three-phase, three-wire ungrounded system require _____ thermal overload units for running overcurrent protection.

30. How is dynamic braking applied to an induction motor? _____

_____

_____

_____

31. How is dynamic braking applied to a synchronous motor? _____

_____

_____

_____

32. What is a megohmmeter? _____

_____

_____

_____

33. Draw a schematic diagram of an across-the-line magnetic switch connected to a three-phase, squirrel-cage induction motor. The magnetic switch has jogging capability. Include in the connection diagram the main relay coil, the pushbutton station with start, jog, and stop pushbuttons, and the maintaining contact.

34. Draw a schematic diagram of a wye-delta starter, complete with pushbutton station, connected to a three-phase motor.

35. On periodic tests, a motor winding suddenly drops to a low resistance value. Testing with a _____ determines this condition.

36. How is a megohmmeter used to measure the insulation resistance of the windings of an ac motor? _____

_____

_____

_____

_____

37. What is a growler? _____

_____

_____

_____

_____

38. Explain how a growler is used to locate a short-circuit condition in a motor winding.

_____

_____

_____

_____

_____

39. For the sleeve bearings of an ac motor, explain how the old oil is removed and the bearings cleaned and lubricated. _____

_____

_____

_____

_____

_____

40. Place the correct answers in each of the spaces provided in the following diagram. Refer to the National Electrical Code.

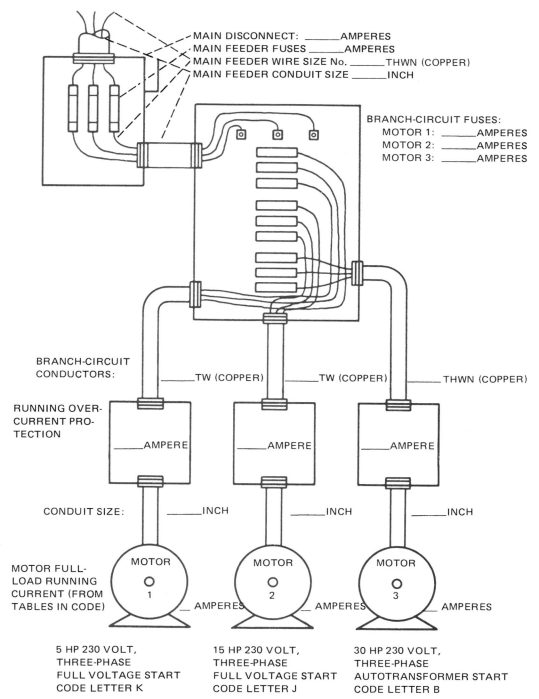

MAIN DISCONNECT: _____AMPERES
MAIN FEEDER FUSES _____AMPERES
MAIN FEEDER WIRE SIZE No. _____THWN (COPPER)
MAIN FEEDER CONDUIT SIZE _____INCH

BRANCH-CIRCUIT FUSES:
  MOTOR 1: _____AMPERES
  MOTOR 2: _____AMPERES
  MOTOR 3: _____AMPERES

BRANCH-CIRCUIT
CONDUCTORS: _____TW (COPPER)   _____TW (COPPER)   _____THWN (COPPER)

RUNNING OVER-
CURRENT PRO-
TECTION   _____AMPERE   _____AMPERE   _____AMPERE

CONDUIT SIZE: _____INCH   _____INCH   _____INCH

MOTOR FULL-
LOAD RUNNING
CURRENT (FROM   _____AMPERES   _____AMPERES   _____AMPERES
TABLES IN CODE)

MOTOR 1   MOTOR 2   MOTOR 3

5 HP 230 VOLT,   15 HP 230 VOLT,   30 HP 230 VOLT,
THREE-PHASE   THREE-PHASE   THREE-PHASE
FULL VOLTAGE START   FULL VOLTAGE START   AUTOTRANSFORMER START
CODE LETTER K   CODE LETTER J   CODE LETTER B

SELSYN UNITS

OBJECTIVES

After studying this unit, the student will be able to

- describe the operation of a simple selsyn system.
- describe the operation of a differential selsyn system.
- list several advantages of selsyn units.

The word *selsyn* is an abbreviation of the words self-synchronous. Selsyn units are special ac motors used primarily in applications requiring remote control. Small selsyn units transmit meter readings or values of various types of electrical and physical quantities to distant points. For example, the captain on the bridge of a ship may adjust the course and speed of the ship; at the same moment, the course and speed changes are transmitted to the engine room by selsyn units. On the engine telegraph system, mechanical positioning of a control transmits electrical angular information to a receiving unit. Similarly, readings of mechanical and electrical conditions in other parts of the ship can be recorded on the bridge by selsyn units. These units are also referred to as *synchros,* and are known by various trade names.

STANDARD SELSYN SYSTEM

A selsyn system consists of two three-phase induction motors. The normally stationary rotors of these induction motors are interconnected so that a manual shift in the rotor position of one machine is accompanied by an electrical rotor shift in the other machine in the same direction and of the same angular displacement as the first unit.

Figure 17-1 shows a simple selsyn system for which the units at the transmitter and receiver are identical. The rotors of these units are two pole and must be excited from the same ac source. The three-phase stator windings are connected to each other by three leads between the transmitter and the receiver units. The rotor of each machine is called the *primary* and the three-phase stator winding of each machine is called the *secondary.* A rotor for a typical selsyn unit is shown in figure 17-2.

When the primary excitation circuit is closed, an ac voltage is impressed on the transmitter and receiver primaries. If both rotors are in the same position with respect to their stators, no movement occurs. If the rotors are not in the same relative position, the freely movable receiver rotor will turn to assume the same position as the transmitter rotor.

If the transmitter rotor is turned, either manually or mechanically, the receiver rotor will follow at the same speed and in the same direction.

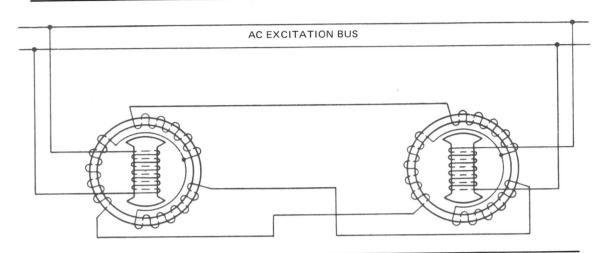

Fig. 17-1   **Diagram of selsyn motors showing interconnected stator and rotor windings connected to excitation source**

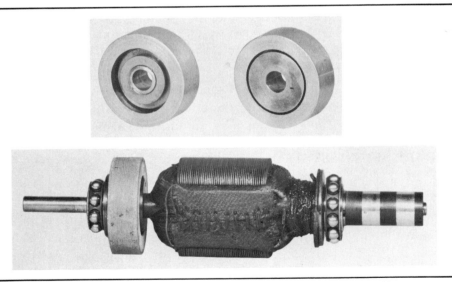

Fig. 17-2   **Wound rotor with oscillation damper and slip rings for selsyn units** *(Photo courtesy of General Electric Company)*

The self-synchronous alignment of the rotors is the result of voltages induced in the secondary windings. Both rotors induce voltages into the three windings of their stators. These voltages vary with the position of the rotors. If the two rotors are in the same relative position, the voltages induced in the transmitter and receiver secondaries will be equal and opposite. In this condition, current will not exist in any part of the secondary circuit.

If the transmitter rotor is moved to another position, the induced voltages of the secondaries are no longer equal and opposite and currents are present in the windings. These currents establish a torque which tends to return the rotors to a synchronous position. Since the receiver rotor is free to move, it makes the adjustment. Any movement of the transmitter rotor is accompanied immediately by an identical movement of the receiver rotor.

DIFFERENTIAL SELSYN SYSTEM

Figure 17-3 is a diagram of the connections of a differential selsyn system consisting of a transmitter, a receiver, and a differential unit. This system produces an angular indication of the receiver. The indication is either the sum or difference of the angles existing at the transmitter and differential selsyns. If two selsyn generators, connected through a differential selsyn, are moved manually to different angles, the differential selsyn will indicate the sum or difference of their angles.

A differential selsyn has a primary winding with three terminals. Otherwise, it closely resembles a standard selsyn unit. The three primary leads of the differential selsyn are brought out to collector rings. The unit has the appearance of a miniature wound-rotor, three-phase induction motor. The unit, however, normally operates as a single-phase transformer.

The voltage distribution in the primary winding of the differential selsyn is the same as that in the secondary winding of the selsyn exciter. If any one of the units is fixed in position and a second unit is displaced by a given angle, then the third unit which is free to rotate will turn through the same angle. The direction of rotation can be reversed by interchanging any pair of leads on either the rotor or stator winding of the differential selsyn.

If any two of the selsyns are rotated simultaneously, the third selsyn will turn through an angle equal to the algebraic sum of the movements of the two selsyns. The algebraic sign of this value depends on the direction of rotation of the rotors of the two selsyns, as well as the phase rotation of their windings.

The excitation current of the differential selsyn is supplied through connections to one or both of the standard selsyns to which the differential selsyn is connected. In general, the excitation current is supplied to the primary winding only. In this case, the selsyn connected to the differential stator supplies this current and must be able to carry the extra load without overheating. A particular type of selsyn, known as an *exciter selsyn*, is used to supply the current. The exciter selsyn can function in the system either as a transmitter or a receiver.

ADVANTAGES OF SELSYN UNITS

Selsyn units are compact and rugged and provide accurate and very reliable readings. Because of the comparatively high torque of the selsyn unit, the indicating pointer does not oscillate as it swings into position. Internal mechanical dampers are used in selsyn receivers to prevent oscillation during the synchronizing procedure and to reduce any tendency of the receiver to operate as a rotor. The operation of the receiver is smooth and continuous and is in agreement with the transmitter. In addition, the response of the receiver to changes in position at the transmitter is very rapid.

In the event of a power failure, the indicator of the receiver resets automatically with the transmitter when power is received. Calibration and time-consuming checks are unnecessary.

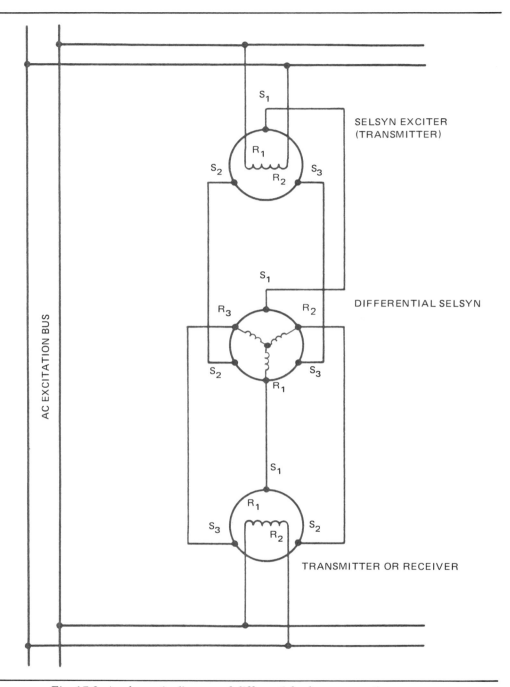

SELSYN EXCITER
(TRANSMITTER)

DIFFERENTIAL SELSYN

TRANSMITTER OR RECEIVER

AC EXCITATION BUS

Fig. 17-3 A schematic diagram of differential selsyn connections

A number of advantages are offered by selsyn systems.

- The indicators are small and compact and can be located where needed.
- The simple installation requires running a few wires and bolting the selsyn units in place.

- Selsyn units can be used to indicate either angular or linear movement.
- Selsyn units control the motion of a device at a distant point by controlling its actuating mechanism.
- One transmitter may be used to operate several receivers simultaneously at several distant points.

ACHIEVEMENT REVIEW

Select the correct answer for each of the following statements and place the corresponding letter in the space provided.

1. Selsyn transmitters and receivers resemble   _____
   a. repulsion-induction motors.
   b. three-phase, two-pole induction motors.
   c. three-phase, four-pole induction motors.
   d. synchronous machines.

2. When the primary excitation circuit is closed, ac voltage is impressed on the   _____
   a. transmitter and receiver primaries.
   b. transmitter rotor and the transmitter stator windings.
   c. transmitter rotor and the receiver stator windings.
   d. stator windings of both instruments.

3. A differential selsyn unit differs from a selsyn transmitter or receiver in that it requires   _____
   a. three-phase power for excitation.
   b. an ac line connection to the stator winding.
   c. dc on the rotor winding.
   d. three connections to the rotor winding.

4. If the rotors of the two selsyn units in a selsyn indicating system are in exactly corresponding positions, the current in the secondary winding is   _____
   a. within quadrature with the primary current.
   b. in phase with the primary current.
   c. zero.
   d. less than the normal primary current.

5. Selsyn units are also referred to as   _____
   a. synchros.   c. wound-rotor motors.
   b. induction motors.   d. all of these.

6. The stator of the transmitter is directly connected to the stator of the receiver unit when a differential is not used.   _____
   a. true   b. false

7. In the transmitter and receiver system, the ac excitation is applied
   to the _____
   a. stator winding.   c. rotor winding only.
   b. stator and the rotor windings. d. none of these.

8. Cite several advantages of a selsyn system. _____

   _____

   _____

   _____

   _____

   _____

SINGLE-PHASE INDUCTION MOTORS

OBJECTIVES

After studying this unit, the student will be able to

- describe the basic operation of the following types of induction motors:

   split-phase motor (both single and dual voltage)
   capacitor start, induction run motor (both single and dual voltage)
   capacitor start, capacitor run motor with one capacitor
   capacitor start, capacitor run motor with two capacitors
   capacitor start, capacitor run motor having an autotransformer with one capacitor

- compare the motors in the listing of objective 1 with regard to starting torque, speed performance, and power factor at the rated load.

The two principal types of single-phase induction motors are the split-phase motor and the capacitor motor. Both types of single-phase induction motors usually have a fractional horsepower rating. The split-phase motor is used to operate such devices as washing machines, small water pumps, oil burners, and other types of small loads not requiring a strong starting torque. The capacitor motor generally is used with devices requiring a strong starting torque, such as refrigerators and compressors. Both types of single-phase induction motors are relatively low in cost, have a rugged construction, and exhibit a good operating performance.

CONSTRUCTION OF A SPLIT-PHASE INDUCTION MOTOR

The split-phase induction motor basically consists of a stator, a rotor, a centrifugal switch located inside the motor, two end shields housing the bearings that support the rotor shaft, and a cast steel frame into which the stator core is pressed. The two end shields are bolted to the cast steel frame. The bearings housed in the end shields keep the rotor centered within the stator so that it will rotate with a minimum of friction and without striking or rubbing the stator core.

The stator for a split-phase motor consists of two windings held in place in the slots of a laminated steel core. The two windings consist of insulated coils distributed and connected to make up two windings spaced 90 electrical degrees apart. One winding is the running winding and the second winding is the starting winding.

The running winding consists of insulated copper wire. It is placed at the bottom of the stator slots. The wire size in the starting winding is smaller than that of the running winding. These coils are placed on top of the running winding coils in the stator slots nearest to the rotor.

Both the starting and running windings are connected in parallel to the single-phase line when the motor is started. After the motor accelerates to a speed equal to approximately two-thirds to three-quarters of the rated speed, the starting winding is disconnected automatically from the line by means of a centrifugal switch.

The rotor for the split-phase motor has the same construction as that of a three-phase, squirrel-cage induction motor. That is, the rotor consists of a cylindrical core assembled from steel laminations. Copper bars are mounted near the surface of the rotor. The bars are brazed or welded to two copper end rings. In some motors, the rotor is a one-piece cast aluminum unit.

Figure 18-1 shows a typical squirrel-cage rotor for a single-phase induction motor. This type of rotor requires little maintenance since there are no windings, brushes, slip rings, or commutators. Note in the figure that the rotor fans are a part of the squirrel-cage rotor assembly. These rotor fans maintain air circulation through the motor to prevent a large increase in the temperature of the windings.

The centrifugal switch is mounted inside the motor. The centrifugal switch disconnects the starting winding after the rotor reaches a predetermined speed, usually two-thirds to three-quarters of the rated speed. The switch consists of a stationary part and a rotating part. The stationary part is mounted on one of the end shields and has two contacts which act like a single-pole, single-throw switch. The rotating part of the centrifugal switch is mounted on the rotor.

A simple diagram of the operation of a centrifugal switch is given in figure 18-2. When the rotor is at a standstill, the pressure of the spring on the fiber ring of the rotating part keeps the contacts closed. When the rotor reaches approximately three-quarters of its rated speed, the centrifugal action of the rotor causes the spring to release its

Fig. 18-1 Cast aluminum squirrel-cage rotor *(Photo courtesy of General Electric Company)*

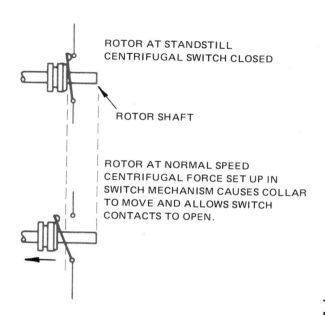

Fig. 18-3 Shaft-mounted centrifugal switch mechanism for integral horsepower single-phase capacitor motor *(Photo courtesy of General Electric Company)*

Fig. 18-2 **Diagram shows the operation of a centrifugal switch**

pressure on the fiber ring and the contacts open. As a result, the starting winding circuit is disconnected from the line. Figure 18-3 is a typical centrifugal switch used with split-phase induction motors.

Principle of Operation

When the circuit to the split-phase induction motor is closed, both the starting and running windings are energized in parallel. Because the running winding consists of a relatively large size of wire, its resistance is low. Recall that the running winding is placed at the bottom of the slots of the stator core. As a result the inductive reactance of this winding is comparatively high due to the mass of iron surrounding it. Since the running winding has a low resistance and a high inductive reactance, the current of the running winding lags behind the voltage.

The starting winding consists of a small size of wire; therefore, its resistance is high. Since the winding is placed near the top of the stator slots, the mass of iron surrounding it is comparatively small and the inductive reactance is low. Therefore, the starting winding has a high resistance and a low inductive reactance. As a result, the current of the starting winding is nearly in phase with the voltage.

The current of the running winding lags the current of the starting winding by approximately 90 electrical degrees. When two currents spaced 90 electrical degrees apart pass through these windings, a rotating magnetic field is developed. This field travels around the inside of the stator core. The speed of the magnetic field is determined using the same procedure given for a three-phase induction motor.

If a split-phase induction motor has four poles on the stator windings and is connected to a single-phase, 60-Hz source, the synchronous speed of the revolving field is:

$$S = \frac{120 \times f}{4}$$

S = synchronous speed

f = frequency in hertz

$$S = \frac{120 \times 60}{4}$$

$$S = 1{,}800 \text{ r/min}$$

As the rotating stator field travels at the synchronous speed, it cuts the copper bars of the rotor and induces voltages in the bars of the squirrel-cage winding. These induced voltages set up currents in the rotor bars. As a result, a rotor field is created which reacts with the stator field to develop the torque which causes the rotor to turn.

As the rotor accelerates to the rated speed, the centrifugal switch disconnects the starting winding from the line. The motor then continues to operate using only the running winding. Figure 18-4 illustrates the connections of the centrifugal switch at the instant the motor starts (switch closed) and when the motor reaches its normal running speed (switch open).

A split-phase motor must have both the starting and running windings energized when the motor is started. The motor resembles a two-phase induction motor because the currents of these two windings are approximately 90 electrical degrees out of phase. The voltage source, however, is single-phase; therefore, the motor is called a split-phase motor because it starts like a two-phase motor from a single-phase line. Once the motor accelerates to a value near its rated speed, it operates on the running winding as a single-phase induction motor.

If the centrifugal switch contacts fail to close when the motor stops, then the starting winding circuit is still open. When the motor circuit is reenergized, the motor will not start. The motor must have both the starting and running windings energized at the instant the motor circuit is closed to create the necessary starting torque. If the motor does not start but simply gives a low humming sound, then the starting winding circuit is open. Either the centrifugal switch contacts are not closed, or there is a break in the coils of the starting windings. **This is an unsafe condition**. The running winding draws excessive current and, therefore, the motor must be removed from the live supply.

If the mechanical load is too great when a split-phase motor is started, or if the terminal voltage applied to the motor is low, then the motor may fail to reach the speed required to operate the centrifugal switch.

The starting winding is designed to operate across the line voltage for a period of only three or four seconds while the motor is accelerating to its rated speed. It is important that the starting winding be disconnected from the line by the centrifugal switch as soon as the motor accelerates to 75 percent of the rated speed. Operation of the motor on its starting winding for more than 60 seconds may char the insulation on the winding or cause the winding to burn out.

To reverse the rotation of the motor, simply interchange the leads of the starting winding (figure 18-5). This causes the direction of the field set up by the stator windings

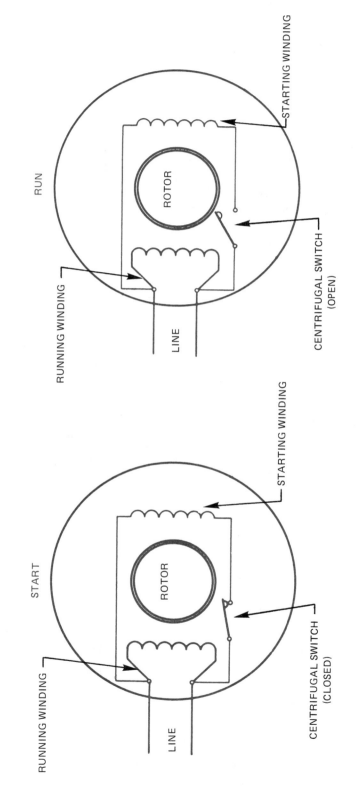

SPLIT-PHASE INDUCTION MOTOR

THE CENTRIFUGAL SWITCH OPENS AT APPROXIMATELY 75 PERCENT OF RATED SPEED

THE STARTING WINDING HAS HIGH RESISTANCE AND LOW INDUCTIVE REACTANCE.
THE RUNNING WINDING HAS LOW RESISTANCE AND HIGH INDUCTIVE REACTANCE.
(PRODUCES 45°–50° PHASE ANGLE FOR STARTING TORQUE.)

Fig. 18-4 Connections of the centrifugal switch at start and at run

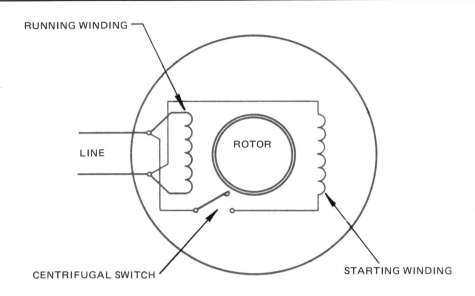

Fig. 18-5 Reversing direction of rotation on split-phase induction motor

to become reversed. As a result, the direction of rotation is reversed. The direction of rotation of the split-phase motor can also be reversed by interchanging the two running winding leads. Normally, the starting winding is used for reversing.

Single-phase motors often have dual-voltage ratings of 115 volts and 230 volts. To obtain these ratings the running winding consists of two sections. Each section of the winding is rated at 115 volts. One section of the running winding is generally marked T_1 and T_2 and the other section is marked T_3 and T_4. If the motor is to be operated on 230 volts, the two 115-volt windings are connected in series across the 230-volt line. If the motor is to be operated on 115 volts, then the two 115-volt windings are connected in parallel across the 115-volt line.

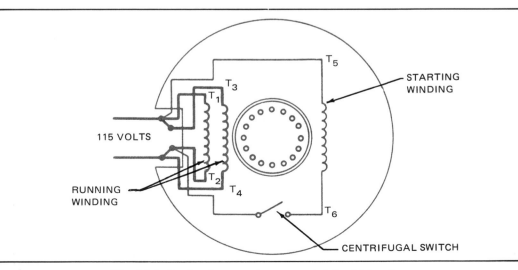

Fig. 18-6 Dual-voltage motor connected for 115 volts

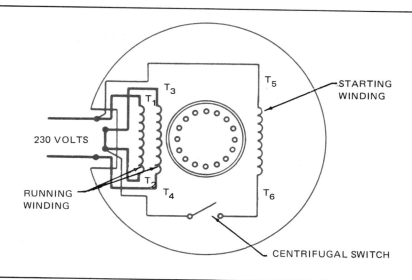

Fig. 18-7 Dual-voltage motor connected for 230 volts

The starting winding, however, consists of only one 115-volt winding. The leads of the starting winding are generally marked T_5 and T_6. If the motor is to be operated on 115 volts, both sections of the running winding are connected in parallel with the starting winding (figure 18-6).

For 230-volt operation, the connection jumpers are changed in the terminal box so that the two 115-volt sections of the running winding are connected in series across the 230-volt line (figure 18-7). Note that the 115-volt starting winding is connected in parallel with one section of the running winding. If the voltage drop across this section of the running winding is 115 volts, then the voltage across the starting winding is also 115 volts.

Some dual-voltage, split-phase motors have a starting winding with two sections as well as a running winding with two sections. The running winding sections are marked T_1 and T_2 for one section and T_3 and T_4 for the other section. One section of the starting winding is marked T_5 and T_6 and the second section of the starting winding is marked T_7 and T_8.

Figure 18-7 shows the winding arrangement for a dual-voltage motor with one starting winding and two running windings. The correct connections for 115-volt operation and for 230-volt operation are given in the table shown in figure 18-8.

The speed regulation of a split-phase induction motor is very good. It has a speed performance from no load to full load that is similar to that of a three-phase, squirrel-cage induction motor. The percent slip on most fractional horsepower split-phase motors is from 4 percent to 6 percent.

The starting torque of the split-phase motor is comparatively poor. The low resistance and high inductive reactance in the running winding circuit, and the high resistance and low inductive reactance in the starting winding circuit cause the two current values to be considerably less than 90 electrical degrees apart. The currents of the starting and running windings in many split-phase motors are only 40 to 50 electrical

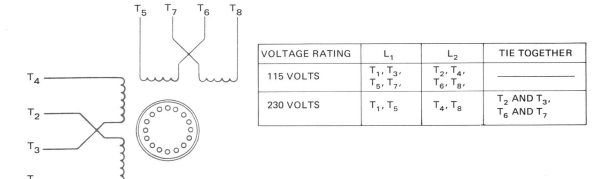

VOLTAGE RATING	L_1	L_2	TIE TOGETHER
115 VOLTS	$T_1, T_3,$ $T_5, T_7,$	$T_2, T_4,$ $T_6, T_8,$	————
230 VOLTS	T_1, T_5	T_4, T_8	T_2 AND $T_3,$ T_6 AND T_7

Fig. 18-8 Winding arrangement for dual-voltage motor with two starting and two running windings

degrees out of phase with each other. As a result, the field set up by these currents does not develop a strong starting torque.

CAPACITOR START, INDUCTION RUN MOTOR

The construction of a capacitor start motor (figure 18-9) is nearly the same as that of a split-phase induction motor. For the capacitor start motor, however, a capacitor is connected in series with the starting windings. The capacitor usually is mounted in a metal casing on top of the motor. The capacitor may be mounted in any convenient external position on the motor frame and, in some cases, may be mounted inside the motor housing. The capacitor provides a higher starting torque than is obtainable with

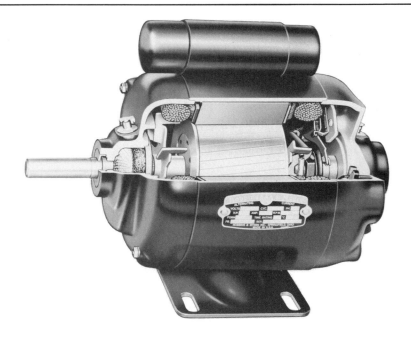

Fig. 18-9 Cutaway view of a fractional horsepower motor *(Photo courtesy of Robbins & Myers Inc.)*

the standard split-phase motor. In addition, the capacitor limits the starting surge of current to a lower value than is developed by the standard split-phase motor.

The capacitor start induction motor is used on refrigeration units, compressors, oil burners, and for small machine equipment, as well as for applications which do not require a strong starting torque.

Principle of Operation

When the capacitor start motor is connected for lower voltage and is started, both the running and starting windings are connected in parallel across the line voltage as the centrifugal switch is closed. The starting winding, however, is connected in series with the capacitor. When the motor reaches a value of 75 percent of its rated speed, the centrifugal switch opens and disconnects the starting winding and the capacitor from the line. The motor then operates as a single-phase induction motor using only the running winding. The capacitor is used to improve the starting torque and does not improve the power factor of the motor.

To produce the necessary starting torque, a revolving magnetic field must be set up by the stator windings. The starting winding current will lead the running winding current by 90 electrical degrees if a capacitor having the correct capacity is connected in series with the starting winding. As a result, the magnetic field developed by the stator windings is almost identical with that of a two-phase induction motor. The starting torque for a capacitor start motor thus is much better than that of a standard split-phase motor.

Defective capacitors are a frequent cause of malfunctions in capacitor start, induction run motors. Some capacitor failures that can occur are:

- the capacitor may short itself out, as evidenced by a lower starting torque.

- the capacitor may be "opened," in which case the starting winding circuits will be open, causing the motor to fail to start.

- the capacitor may short circuit and cause the fuse protection for the branch motor circuit to blow. If the fuse ratings are quite high and do not interrupt the power supply to the motor soon enough, the starting winding may burn out.

- starting capacitors can short circuit if the motor is turned on and off many times in a short period of time. To prevent capacitor failures, many motor manufacturers recommend that a capacitor start motor be started no more than 20 times per hour. Therefore, this type of motor is used only in those applications where there are relatively few starts in a short time period.

The speed performance of a capacitor start motor is very good. The increase in percent slip from a no-load condition to full load is from 4 percent to 6 percent. The speed performance then is the same as that of a standard split-phase motor.

The leads of the starting winding circuit are interchanged to reverse the direction of rotation of a capacitor start motor. As a result, the direction of rotation of the magnetic field developed by the stator windings reverses in the stator core, and the rotation of the rotor is reversed.

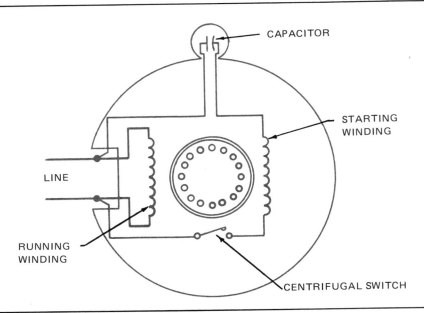

Fig. 18-10 Connections for a capacitor start, induction run motor

Figure 18-10 is a diagram of the circuit connections of a capacitor start motor before the starting winding leads are interchanged to reverse the direction of rotation of the rotor. The diagram in figure 18-11 shows the circuit connections of the motor after the starting winding leads are interchanged to reverse the direction of rotation.

A second method of changing the direction of rotation of a capacitor start motor is to interchange the two running winding leads. However, this method is seldom used.

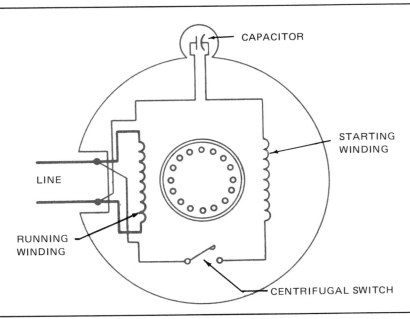

Fig. 18-11 Connections for reversing a capacitor start, induction run motor

Capacitor start, induction run motors often have dual-voltage ratings of 115 volts and 230 volts. The connections for a capacitor start motor are the same as those for split-phase induction motors.

CAPACITOR START, CAPACITOR RUN MOTOR

The capacitor start, capacitor run motor is similar to the capacitor start, induction run motor, except that the starting winding and capacitor are connected in the circuit at all times. This motor has a very good starting torque. The power factor at the rated load is nearly 100 percent or unity due to the fact that a capacitor is used in the motor at all times.

There are several different designs for this type of motor. One type of capacitor start, capacitor run motor has two stator windings which are spaced 90 electrical degrees apart. The main or running winding is connected directly across the rated line voltage. A capacitor is connected in series with the starting winding and this series combination also is connected across the rated line voltage. A centrifugal switch is not used because the starting winding is energized through the entire operating period of the motor.

Figure 18-12 illustrates the internal connections for a capacitor start, capacitor run motor using one value of capacitance.

To reverse the rotation of this motor, the lead connections of the starting winding must be interchanged. This type of capacitor start, capacitor run motor is quiet in operation and is used on oil burners, fans, and small wood working and metal working machines.

A second type of capacitor start, capacitor run motor has two capacitors. Figure 18-13 is a diagram of the internal connections of the motor. At the instant the motor is started, the two capacitors are in parallel. When the motor reaches 75 percent of the rated speed, the centrifugal switch disconnects the larger capacitor. The motor then operates with the smaller capacitor only connected in series with the starting winding.

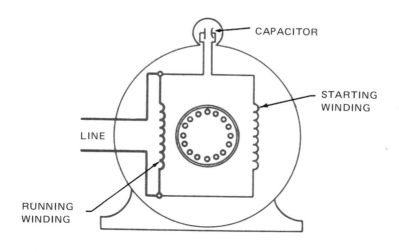

Fig. 18-12 Connections for a capacitor start, capacitor run motor

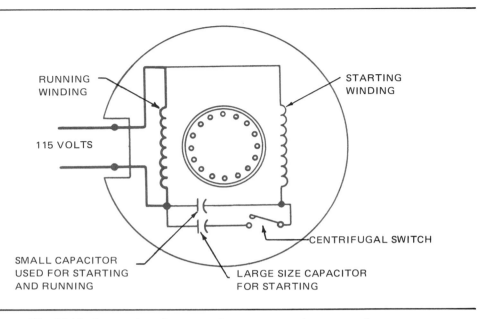

Fig. 18-13 Connections for a capacitor start, capacitor run motor

This type of motor has a very good starting torque, good speed regulation, and a power factor of nearly 100 percent at rated load. Applications for this type of motor include furnace stokers, refrigerator units, and compressors.

A third type of capacitor start, capacitor run motor has an autotransformer with one capacitor. This motor has a high starting torque and a high operating power factor. Figure 18-14 is a diagram of the internal connections for this motor. When the motor is started, the centrifugal switch connects winding 2 to point A on the tapped

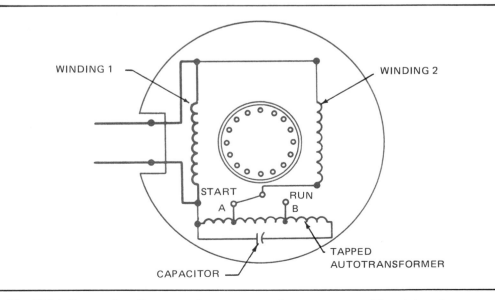

Fig. 18-14 Connections for a capacitor start, capacitor run motor with autotransformer

autotransformer. As the capacitor is connected across the maximum transformer turns, it receives maximum voltage output on start. The capacitor thus is connected across a value of approximately 500 volts. As a result, there is a high value of leading current in winding 2 and a strong starting torque is developed.

When the motor reaches approximately 75 percent of the rated speed, the centrifugal switch disconnects the starting winding from point A and reconnects this winding to point B on the autotransformer. Less voltage is applied to the capacitor, but the motor operates with both windings energized. Thus, the capacitor maintains a power factor near unity at the rated load.

The starting torque of this motor is very good and the speed regulation is satisfactory. Applications requiring these characteristics include refrigerators and compressors.

NATIONAL ELECTRICAL CODE REGULATIONS

Section 430-32(b)(1) of the National Electrical Code states that any motor of one horsepower or less which is manually started and is within sight from the starter location, shall be considered as protected against overload by the overcurrent device protecting the conductors of the branch circuit. This branch overcurrent device shall not be larger than specified in *Article 430, Part D (Motor Branch Circuit, Short-Circuit and Ground-Fault Protection)*. An exception is that any such motor may be used at 125 volts or less on a branch circuit protected at not over 20 amperes.

A distance of more than 50 feet is considered to be out of sight from the starter location. *Section 430-32(c)* covers motors, one horsepower or less, automatically started, which are out of sight from the starter location or permanently installed.

Section 430-32(c)(1) states that any motor of one horsepower or less which is started automatically shall have a separate overcurrent device which is responsive to the motor current. This overload unit shall be set at not more than 125 percent of the full-load current rating of the motor for motors marked to have a temperature rise not over 40 degrees Celsius or with a service factor not less than 1.15, and at not more than 115 percent for all other types of motors.

ACHIEVEMENT REVIEW

1. List the essential parts of a split-phase induction motor. _____

   _____

   _____

   _____

2. What happens if the centrifugal switch contacts fail to reclose when the motor stops? _____

   _____

   _____

   _____

3. Explain how the direction of rotation of a split-phase induction motor is reversed.

_____

_____

_____

_____

4. A split-phase induction motor has a dual-voltage rating of 115/230 volts. The motor has two running windings, each of which is rated at 115 volts, and one starting winding rated at 115 volts. Draw a schematic diagram of this split-phase induction motor connected for 230-volt operation.

5. Draw a schematic connection diagram of the split-phase induction motor in question 4 connected for 115-volt operation.

6. A split-phase induction motor has a dual-voltage rating of 115/230 volts. The motor has two running windings, each of which is rated at 115 volts. In addition, there are two starting windings and each of these windings is rated at 115 volts. Draw a schematic connection diagram of this split-phase induction motor connected for 230-volt operation.

7. What is the primary difference between a split-phase induction motor and a capacitor start, induction run motor? _____

_____

_____

_____

8. If the centrifugal switch fails to open as a split-phase motor accelerates to its rated speed, what will happen to the starting winding? _____

_____

_____

_____

9. What is one limitation of a capacitor start, induction run motor? _____

_____

_____

_____

10. Insert the correct word or phrase to complete each of the following statements.

   a. A motor of one horsepower or less which is manually started and which is within sight of the starter location is considered to be protected by the _____

   _____ .

   b. A motor of one horsepower or less which is manually started is considered within sight of the starter location if the distance is no greater than _____

   _____ .

   c. The capacitor used with a capacitor start, induction run motor is used only to improve the _____ .

   d. A capacitor start, induction run motor has a better starting torque than the

   _____ .

REPULSION MOTORS

OBJECTIVES

After studying this unit, the student will be able to

- describe the basic steps in the operation of the following types of motors:
    repulsion motor
    repulsion start, induction run motor
    repulsion-induction motor
- state the basic construction differences among the motors listed in objective 1.
- compare the motors listed in objective 1 with regard to starting torque and speed performance.

Repulsion-type motors are divided into three distinct classifications: the repulsion motor; the repulsion start, induction run motor; and the repulsion-induction motor. Although these motors are similar in name, they differ in construction, operating characteristics, and industrial applications.

REPULSION MOTOR

A repulsion motor basically consists of the following parts:

Laminated stator core with one winding. This winding is similar to the main or running winding of a split-phase motor. The stator usually is wound with four, six, or eight poles.

Rotor consisting of a slotted core into which a winding is placed. The rotor is similar in construction to the armature of a dc motor. Thus, the rotor is called an armature. The coils which make up this armature winding are connected to a commutator. The commutator has segments or bars parallel to the armature shaft.

Carbon brushes contacting with the commutator surface. The brushes are held in place by a brush holder assembly mounted on one of the end shields. The brushes are connected together by heavy copper jumpers. The brush holder assembly may be moved so that the brushes can make contact with the commutator surface at different points to obtain the correct rotation and maximum torque output. There are two types of brush arrangements:

1. Brush riding — the brushes are in contact with the commutator surface at all times.
2. Brush lifting — the brushes lift at approximately 75 percent of the rotor speed.

Two cast steel end shields. These shields house the motor bearings and are secured to the motor frame.

Two bearings supporting the armature shaft. The bearings center the armature with respect to the stator core and windings. The bearings may be sleeve bearings or ball bearing units.

Cast steel frame into which the stator core is pressed.

Operation of a Repulsion Motor

The connection of the stator winding of a repulsion motor to a single-phase line causes a field to be developed by the current in the stator windings. This stator field induces a voltage and a resultant current in the rotor windings. If the brushes are placed in the proper position on the commutator segments, the current in the armature windings will set up proper magnetic poles in the armature.

These armature field poles have a set relationship to the stator field poles. That is, the magnetic poles developed in the armature are set off from the field poles of the stator winding by about 15 electrical degrees. Furthermore, since the instantaneous polarity of the rotor poles is the same as that of the adjacent stator poles, the repulsion torque created causes the rotation of the motor armature.

The three diagrams of figure 19-1 show the importance of the brushes being in the proper position to develop maximum torque. In figure 19-1A, no torque is developed when the brushes are placed at right angles to the stator poles. This is due to the fact that the equal induced voltages in the two halves of the armature winding oppose each other at the connection between the two sets of brushes. Since there is no current in the windings, flux is not developed by the armature windings.

In figure 19-1B, the brushes are in a position directly under the center of the stator poles. A heavy current exists in the armature windings with the brushes in this position, but there is still no torque. The heavy current in the armature windings sets up poles in the armature. However, these poles are centered with the stator poles and a torque is not created either in a clockwise or counterclockwise direction.

In figure 19-1C, the brushes have shifted from the center of the stator poles 15 electrical degrees in a counterclockwise direction. Thus, magnetic poles of like polarity are set up in the armature. These poles are 15 electrical degrees in a counterclockwise direction from the stator pole centers. A repulsion torque is created between the stator and the rotor field poles of like polarity. The torque causes the armature to rotate in a counterclockwise direction. A repulsion machine has a high starting torque, with a small starting current, and a rapidly decreasing speed with an increasing load.

The direction of rotation of a repulsion motor is reversed if the brushes are shifted 15 electrical degrees from the stator field pole centers in a clockwise direction, figure 19-2. As a result, magnetic poles of like polarity are set up in the armature. These poles are 15 electrical degrees in a clockwise direction from the stator pole centers. Repulsion motors are used principally for constant-torque applications, such as printing-press drives, fans, and blowers.

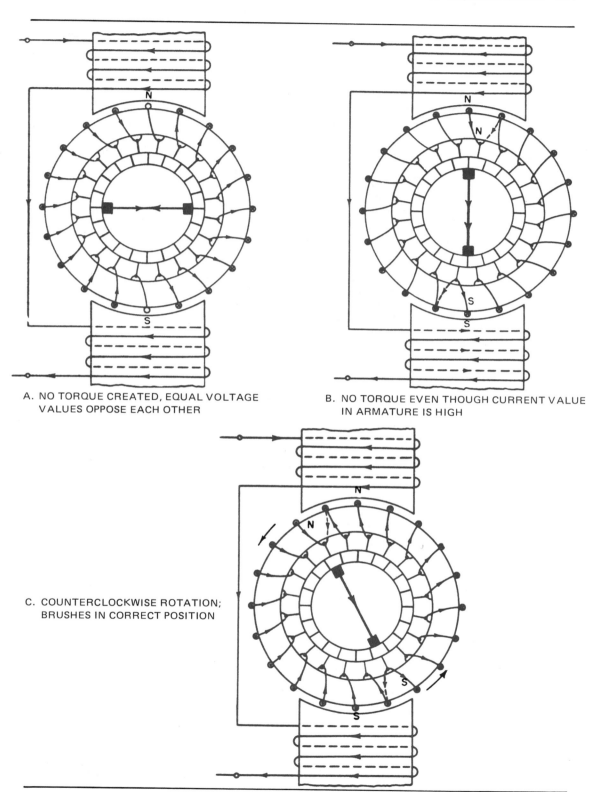

A. NO TORQUE CREATED, EQUAL VOLTAGE
   VALUES OPPOSE EACH OTHER

B. NO TORQUE EVEN THOUGH CURRENT VALUE
   IN ARMATURE IS HIGH

C. COUNTERCLOCKWISE ROTATION;
   BRUSHES IN CORRECT POSITION

Fig. 19-1 Repulsion motor operation

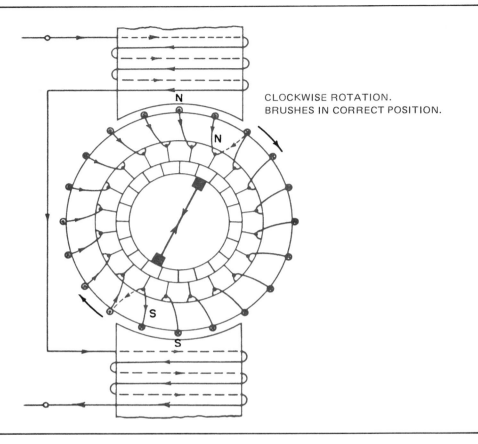

CLOCKWISE ROTATION.
BRUSHES IN CORRECT POSITION.

Fig. 19-2 Reversing the direction of rotation of a repulsion motor

REPULSION START, INDUCTION RUN MOTOR

A second type of repulsion motor is the repulsion start, induction run motor. In this type of motor, the brushes contact the commutator at all times. The commutator of this motor is the more conventional axial form.

A repulsion start, induction run motor consists basically of the following parts.

Laminated stator core. This core has one winding which is similar to the main or running winding of a split-phase motor.

Rotor consisting of a slotted core into which a winding is placed. The coils which make up the winding are connected to a commutator. The rotor core and winding are similar to the armature of a dc motor. Thus, the rotor is called an armature.

Centrifugal device.
   a. In the brush-lifting type of motor, there is a centrifugal device which lifts the brushes from the commutator surface at 75 percent of the rated speed. This device consists of governor weights, a short-circuiting necklace, a spring barrel, spring, push rods, brush holders, and brushes. Although high in first cost, this device does save wear and tear on brushes, and runs quietly.

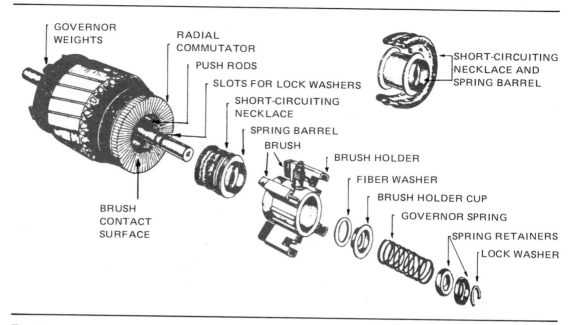

Fig. 19-3 An exploded view of a radial commutator and centrifugal brush-lifting device for a repulsion start, induction run motor

Figure 19-3 is an exploded view of the armature, radial commutator, and centrifugal device of the brush-lifting type of repulsion start, induction run motor.

b. The brush-riding type of motor also contains a centrifugal device which operates at 75 percent of the rated speed. This device consists of governor weights, a short-circuiting necklace, and a spring barrel. The commutator segments are short circuited by this device, but the brushes and brush holders are not lifted from the commutator surface.

Commutator. The brush-lifting type of motor has a radial-type commutator (figure 19-3). The brush-riding type of motor has an axial commutator (figure 19-4).

Brush holder assembly.

a. The brush holder assembly for the brush-lifting type of motor is arranged so that the centrifugal device can lift the brush holders and brushes clear of the commutator surface.

b. The brush holder assembly for the brush-riding type of motor is the same as that of a repulsion motor.

End shields, bearings, and motor frame. The parts are the same as those of a repulsion motor.

Operation of the Centrifugal Mechanism

Refer to figure 19-3 to identify the components of the centrifugal mechanism. The operation of this device consists of the following steps. As the push rods of the

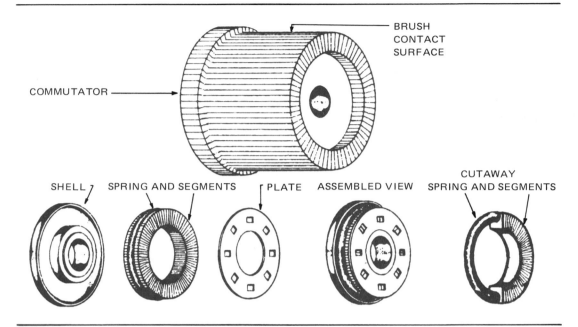

Fig. 19-4   An exploded view of a short-circuiting device for a brush-riding, repulsion start, induction run motor

centrifugal device move forward, they push the spring barrel forward. This allows the short-circuiting necklace to make contact with the radial commutator bars which thus are all short circuited. At the same time, the brush holders and brushes are moved from the commutator surface. As a result, there is no unnecessary wear on the brushes and the commutator surface and there are no objectionable noises caused by the brushes riding on the radial commutator surface.

The short-circuiting action of the governor mechanism and the commutator segments converts the armature to a form of squirrel-cage rotor and the motor operates as a single-phase induction motor. In other words, the motor starts as a repulsion motor and runs as an induction motor.

In the brush-riding type of motor, an axial commutator is used. The centrifugal mechanism (figure 19-4) consists of a number of copper segments which are held in place by a spring. This device is placed next to the commutator. When the rotor reaches 75 percent of the rated speed, the centrifugal device short circuits the commutator segments. The motor then will continue to operate as an induction motor.

Operation of a Repulsion Start, Induction Run Motor

The starting torque is good for either the brush-lifting type or the brush-riding type of repulsion start, induction run motor. Furthermore, the speed performance of both types of motors is very good since they operate as single-phase induction motors.

Because of the excellent starting and running characteristics for both types of repulsion start, induction run motors, they are used for a variety of industrial applications, including commercial refrigerators, compressors, and pumps.

Fig. 19-5 Schematic diagram symbol of a repulsion start, induction run motor and a repulsion motor

The direction of rotation for a repulsion start, induction run motor is changed in the same manner as that for a repulsion motor, that is, by shifting the brushes past the stator pole center 15 electrical degrees.

The symbol in figure 19-5 represents both a repulsion start, induction run motor and a repulsion motor.

Many repulsion start, induction run motors are designed to operate on 115 volts or 230 volts. These dual-voltage motors contain two stator windings. For 115-volt operation, the stator windings are connected in parallel; for 230-volt operation, the stator windings are connected in series. The diagram and photo in figure 19-6A and B represent a dual-voltage, repulsion start, induction run motor. The connection table in the figure shows how the leads of the motor are connected for either 115-volt operation or 230-volt operation. These connections also can be used for dual-voltage repulsion motors.

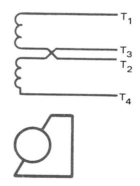

	L_1	L_2	TIE TOGETHER
LOW VOLTAGE	T_1 T_3	T_2 T_4	——————
HIGH VOLTAGE	T_1	T_4	T_2 T_3

Fig. 19-6 A) Single-phase repulsion induction motor *(Photo courtesy of Peerless-Winsmith Inc.)*
B) Schematic diagram of a dual-voltage, repulsion start, induction run motor

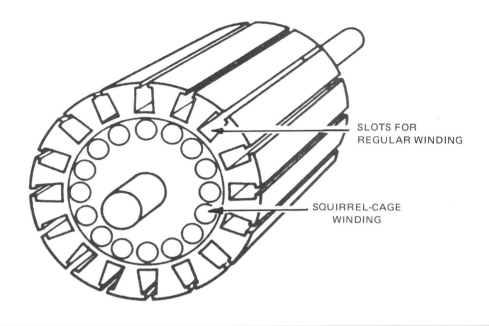

SLOTS FOR
REGULAR WINDING

SQUIRREL-CAGE
WINDING

Fig. 19-7 An armature of a repulsion-induction motor

REPULSION-INDUCTION MOTOR

The operating characteristics of a repulsion-induction motor are similar to those of the repulsion start, induction run motor. However, the repulsion-induction motor has no centrifugal mechanism. It has the same type of armature and commutator as the repulsion motor, but it has a squirrel-cage winding beneath the slots of the armature.

Figure 19-7 shows a repulsion-induction motor armature with a squirrel-cage winding. One advantage of this type of motor is that it has no centrifugal device requiring maintenance. The repulsion-induction motor has a very good starting torque since it starts as a repulsion motor. At start up, the repulsion winding predominates; but, as the motor speed increases, the squirrel-cage winding becomes predominant. The transition from repulsion to induction operation is smooth since no switching device is used. In addition, the repulsion-induction motor has a fairly constant speed regulation from no load to full load because of the squirrel-cage winding. The torque-speed performance of a repulsion-induction motor is similar to that of a dc compound motor.

A repulsion-induction motor can be operated on either 115 volts or 230 volts. The stator winding has two sections which are connected in parallel for 115-volt operation, and in series for 230-volt operation. The markings of the motor terminals and the connection arrangement of the leads is the same as in a repulsion start, induction run motor.

The symbol in figure 19-5 also represents a repulsion-induction motor (as well as a repulsion start, induction run motor and a repulsion motor.)

NATIONAL ELECTRICAL CODE REGULATIONS

Regulations for the motor branch circuit overcurrent protection, motor running overcurrent protection, and wire sizes for motor circuits are given in *Article 430* of the National Electrical Code. Refer also to *Example 8, Chapter 9* of the Code.

ACHIEVEMENT REVIEW

1. What is a repulsion motor, and how is rotation produced? _____

_____

_____

_____

_____

_____

2. Name one application of a repulsion motor. _____

_____

_____

3. Describe the operation of a repulsion start, induction run motor. _____

_____

_____

_____

_____

4. Explain the difference between the brush-lifting type of repulsion start, induction run motor and the brush-riding type of repulsion start, induction run motor.

_____

_____

_____

_____

5. A 2-hp, 230-volt, 12-ampere, single-phase repulsion start, induction run motor is connected directly across the rated line voltage.

a. Determine the overcurrent protection for the branch circuit feeding this motor.

_____

_____

b. Determine the running overcurrent protection for this motor.

_____

_____

6. What size wire is used for the branch circuit feeding the motor given in question 5? _____

_____

7. Describe the construction of a repulsion-induction motor. _____

_____

_____

_____

8. What is one advantage to the use of the repulsion-induction motor as compared to to the repulsion start, induction run motor? _____

_____

_____

9. Explain how the direction of rotation is changed for any one of the three types of single-phase repulsion motors covered in this unit. _____

_____

_____

10. Insert the correct word or phrase to complete each of the following statements.

a. A repulsion-induction motor has a good _____
and a fairly good _____ .

b. A repulsion motor has a high starting torque and its speed rapidly decreases with _____ .

c. The centrifugal short-circuiting device on a repulsion start, induction run motor operates at approximately _____ of the rated speed.

d. Both the repulsion start, induction run motor, and the repulsion-induction motor operate as _____ after they have accelerated to rated speed.

ALTERNATING-CURRENT SERIES MOTORS

OBJECTIVES

After studying this unit, the student will be able to

- describe the basic operation of a universal motor.
- explain how a single-field compensated universal motor operates.
- explain how a two-field compensated universal motor operates.
- describe two ways in which universal motors are compensated for excessive armature reaction under load.
- state the reasons why dc motors fail to operate satisfactorily from an ac source.

The electrician may consider a typical dc series motor or a dc shunt motor for operation on ac power supplies. It appears that such operation is possible since reversing the line terminals to a dc motor reverses the current and magnetic flux in both the field and armature circuits. As a result, the net torque of the motor operating from an ac source is in the same direction.

However, the operation of a dc shunt motor from an ac source is impractical because the high inductance of the shunt field causes the field current and the field flux to lag the line voltage by almost 90 degrees. The resulting torque is very low.

A dc series motor also fails to operate satisfactorily from an ac source because of the excessive heat developed by eddy currents in the field poles. In addition, there is an excessive voltage drop across the series field windings due to high reactance.

To reduce the eddy currents, the field poles can be laminated. To reduce the voltage loss across the field poles to a minimum, a small number of field turns can be used on a low reluctance core operated at low flux density. A motor with these revisions operates on either ac or dc and is known as a universal motor. Universal motors in small fractional horsepower sizes are used in household appliances and portable power tools.

CONCENTRATED FIELD UNIVERSAL MOTOR

A concentrated field universal motor usually is a salient-pole motor with two poles and a winding of relatively few turns. The poles and winding are connected to give opposite magnetic polarity. A field yoke of this type of motor is shown in figure 20-1.

DISTRIBUTED FIELD UNIVERSAL MOTORS

The two types of distributed field universal motors are: the single-field compensated motor and the two-field compensated motor. The major parts of a distributed field universal motor are shown in figure 20-2.

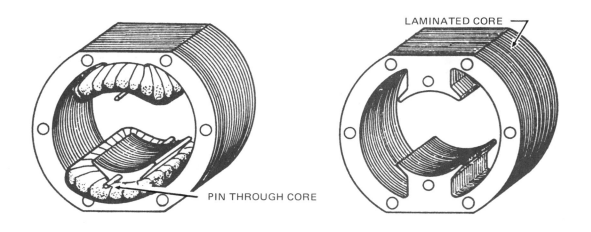

Fig. 20-1 Field core of a two-pole universal motor

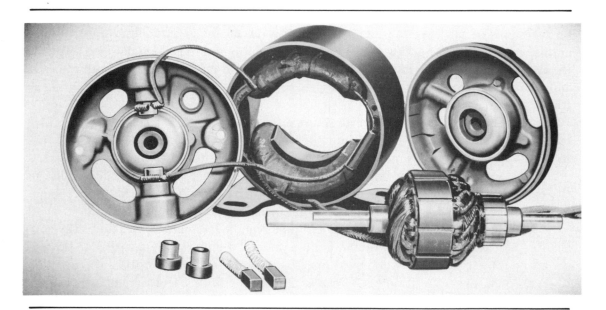

Fig. 20-2 Disassembled view of a distributed field universal motor *(Photo courtesy of Robbins & Myers Inc.)*

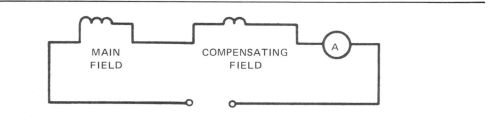

Fig. 20-3 Schematic diagram of a compensated universal motor

The field windings of a two-pole, single-field compensated motor resemble the stator winding of a two-pole, split-phase ac motor. A two-field compensated motor has a stator containing a main winding and a compensating winding spaced 90 electrical degrees apart. The compensating winding reduces the reactance voltage developed in the armature by the alternating flux when the motor operates from an ac source. Figure 20-3 is the schematic diagram of a compensated universal motor.

THE ARMATURE

The armature of a typical universal motor resembles the armature of a typical dc motor except that a universal motor armature is slightly larger for the same horsepower output.

CONSTRUCTION FEATURES OF UNIVERSAL MOTORS

The frames of universal motors are made of aluminum, cast iron, and rolled steel. The field poles are generally bolted to the frame. Field cores consist of laminations pressed together and held by bolts. The armature core is also laminated and has a typical commutator and brushes. End plates resemble those of other motors except that in many universal motors one end plate is cast as part of the frame. Both ball and sleeve bearings are used in universal motors.

SPEED CONTROL

Universal motors operate at approximately the same speed on dc or single-phase ac. Since these motors are series wound, they will operate at excessive speed at a no-load condition. As a result, they usually are permanently connected to the devices being driven. Universal motors are speed regulated by inserting resistance in series with the motor. The resistance may be tapped resistors, rheostats, or tapped nichrome wire coils wound over a single field pole. In addition, speed may be controlled by varying the inductance through taps on one of the field poles. Gear boxes are also used.

DIRECTION OF ROTATION

The direction of rotation of any series wound motor can be reversed by changing the direction of the current in either the field or the armature circuit. Universal motors are sensitive to brush position and severe arcing at the brushes will result from changing the direction of rotation without shifting the brushes to the neutral (sparkless) plane.

CONDUCTIVE COMPENSATION

Ac motors rated at more than 1/2 horsepower are used to drive loads requiring a high starting torque. Two methods are used to compensate for excessive armature reaction under load. In the conductively compensated type of motor, an additional compensating winding is placed in slots cut directly into the pole faces. The strength of this field increases with an increase in load current and thus minimizes the distortion of the main field flux by the armature flux. The compensating winding is connected

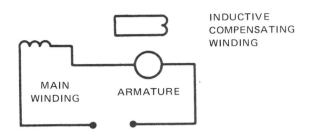

Fig. 20-4 Connections for an inductively compensated universal motor

in series with the series field winding and the armature, as shown in figure 20-3. Although conductively compensated motors have a high starting torque, the speed regulation is poor. A wide range of speed control is possible with the use of resistor-type starter-controllers.

INDUCTIVE COMPENSATION

Armature reaction in ac series motors also may be compensated with an inductively coupled winding which acts as a short-circuited secondary winding of a transformer. This winding is placed so that it links the cross-magnetizing flux of the armature which acts as the primary winding of a transformer. Figure 20-4 is the schematic diagram of an inductively compensated universal motor. Since the magnetomotive force of the secondary is nearly opposite in phase and equal in magnitude to the primary magnetomotive force, the compensating winding flux nearly neutralizes the armature cross flux. This type of motor cannot be used on dc current. Because of its dependency on induction, the operating characteristics of an inductively compensated motor are very similar to those of the conductively compensated motor.

ACHIEVEMENT REVIEW

A. Completely answer the following questions.

1. a. Describe the basic differences in construction between the concentrated field and the distributed field types of universal motors. b. Draw the schematic diagram for each type of motor.

a. _____

_____

_____

_____

b.

2. What is the function of the compensating winding in a two-field compensated universal motor? _____
_____
_____

3. Describe three methods of controlling the speed of universal motors. _____
_____
_____
_____

4. Why does a universal motor spark excessively at the commutator if its direction of rotation is reversed? _____
_____
_____

5. A dc series motor operates unsatisfactorily on ac. What are the primary reasons for this fact? _____
_____
_____
_____

B. Select the correct answer for each of the following statements and place the corresponding letter in the space provided.

6. The operation of a dc shunt motor from an ac source is impractical because _____
   a. too much torque is developed at startup.
   b. the starting current is too high.
   c. the shunt field inductance is too high.
   d. the shunt field inductance is too low.

7. A series dc motor fails to operate satisfactorily on ac due to   _____
   a. eddy currents and high field voltage drop.
   b. excessive heat and low field voltage drop.
   c. low reactance of the armature and field.
   d. high armature reluctance and low field reactance.

8. The frames of universal motors are made of   _____
   a. rolled steel.   c. aluminum.
   b. cast iron.   d. all of these.

9. A compensating winding   _____
   a. increases the reactance in the armature on ac.
   b. reduces the reactance in the armature on ac.
   c. reduces the reactance in the armature on dc.
   d. increases the reactance in the armature on dc.

10. After changing the direction of rotation of a universal motor, the   _____
   a. brushes must be rotated for sparkless commutation.
   b. field connections must be shifted.
   c. field reactance must be decreased.
   d. field reactance must be increased.

SUMMARY REVIEW OF UNITS 17-20

OBJECTIVE

- To give the student an opportunity to evaluate the knowledge and understanding acquired in the study of the previous four units.

1. List four applications for selsyn units.

   a. _____

   b. _____

   c. _____

   d. _____

2. Draw a schematic wiring diagram of a selsyn system consisting of a transmitter and one selsyn receiver unit.

3. Explain the operation of a simple selsyn system consisting of one transmitter and one selsyn receiver unit. _____

   _____

   _____

   _____

   _____

   _____

4. Explain the purpose of a differential selsyn unit. _____

_____

_____

_____

_____

5. Draw a schematic wiring diagram of a selsyn system having one selsyn exciter unit, a differential selsyn unit, and one selsyn receiver unit.

6. Insert the correct word or phrase to complete each of the following statements.

   a. Whenever the rotor of the _____ is out of alignment with the rotor of the receiver selsyn, currents are present in the stator windings.

   b. A selsyn receiver will rotate continuously if the transmitter rotor is driven at _____ speed.

   c. The rotor units of the transmitter and receiver selsyns must be excited from the same _____ .

   d. The use of a mechanical damper on the rotor of selsyn receivers minimizes any tendency of the receiver to _____ .

7. List four applications for a split-phase induction motor.

   a. _____

   b. _____

   c. _____

   d. _____

8. What are the basic parts of a split-phase induction motor? _____

_____

_____

_____

9. Explain how the direction of rotation of a split-phase induction motor is reversed.

_____

_____

_____

10. What happens if the centrifugal switch contacts fail to reclose when a split-phase motor is stopped? _____

_____

11. A split-phase induction motor has a dual-voltage rating of 115/230 volts. The motor has two running windings, each of which is rated at 115 volts. The motor also has two starting windings, each of which is rated at 115 volts. Draw a schematic connection diagram of this split-phase induction motor connected for 230 volts.

12. What is the basic difference between a split-phase induction motor and a capacitor start, induction run motor? _____

_____

_____

_____

_____

_____

_____

13. If the centrifugal switch fails to open as a split-phase motor accelerates to the rated speed, what happens to the starting winding? _____

_____

14. If the centrifugal switch on a capacitor start, induction run motor fails to open as the motor accelerates to the rated speed, what may happen in the starting winding circuit? _____

_____

15. What is one limitation of a capacitor start, induction run motor? _____

_____

_____

16. What is the basic difference between a capacitor start, induction run motor and a capacitor start, capacitor run motor? _____

_____

_____

_____

_____

17. List three types of capacitor start, capacitor run motors.

    a. _____

    b. _____

    c. _____

18. Insert the correct word or phrase to complete each of the following statements.

    a. The capacitor in series with the starting winding of a capacitor start, induction run motor improves the _____ torque of the motor.

    b. A split-phase induction motor has good speed regulation but _____ starting torque characteristics.

    c. A capacitor start, capacitor run motor has practically _____ power factor when operating at full load.

    d. A capacitor start, capacitor run motor has _____ starting torque and _____ speed regulation.

    e. A capacitor start, induction run motor has _____ speed regulation.

19. Insert the correct word or phrase to complete each of the following statements.

    a. The capacitor used with a capacitor start, induction run motor is used only for the purpose of improving the _____ of the motor.

    b. The capacitors used with a capacitor start, capacitor run motor are used to improve _____ .

    c. A motor of one horsepower or less which is manually started and which is within sight of the starter location, provided the distance is no greater than 50 feet, is considered protected by the _____ .

    d. A motor of one horsepower or less which is manually operated but more than 50 feet from the starter location shall have a _____
_____ .

    e. Where separate overcurrent devices are required for motors, they shall not be set at more than _____ percent of the motor nameplate full-load current rating for motors marked to have a temperature rise not over _____ for motors with a marked service factor of 1.15, and at not more than _____ percent for all other types of motors.

20. Where is a repulsion motor used? _____

_____

_____

_____

_____

21. What are the two types of repulsion start, induction run motors?

    a. _____

    b. _____

22. Where are repulsion start, induction run motors used? _____

_____

_____

23. A 3-hp, 230-volt, 17-ampere, single-phase repulsion start, induction run motor is connected directly across rated line voltage.

    a. Determine the overcurrent protection for the branch circuit feeding this motor.

_____

_____

    b. Determine the running overcurrent protection to use with this motor. _____

_____

_____

24. What size copper wire (Type THHN) is used for the branch circuit feeding the motor in question 23? _____

_____

_____

25. What is a repulsion-induction motor? _____

_____

_____

_____

26. What is one advantage to the use of the repulsion-induction motor as compared with the repulsion start, induction run motor? _____

_____

_____

27. Insert the correct word or phrase to complete each of the following statements.

   a. A repulsion motor has good _____ but poor _____ .

   b. The speed of a repulsion motor can be controlled by changing the _____ .

   c. Both the brush-riding and the brush-lifting types of repulsion start, induction run motors operate as _____ after they have accelerated to rated speed.

   d. The repulsion-induction motor has good _____ and relatively good _____ .

28. Explain how the direction of rotation is changed on any one of the three types of single-phase repulsion motors. _____

_____

_____

29. What is a universal motor? _____

_____

_____

30. Draw a schematic diagram of a conductively compensated series motor.

31. Draw a schematic diagram of an inductively compensated series motor.

32. In what way is an inductively compensated series motor different from a conductively compensated series motor? _____

_____

_____

_____

_____

33. Explain how the direction of rotation is reversed for a conductively compensated series motor. _____

_____

_____

34. What is the purpose of a compensating winding in an ac series motor? _____

_____

_____

_____

_____

_____

35. A universal motor can be operated on   _____
    a. ac power only.
    b. dc power only.
    c. ac or dc power.

36. A conductively compensated series motor can be operated on   _____
    a. ac power only.
    b. dc power only.
    c. ac or dc power.

37. An inductively compensated series motor can be operated on
    either ac or dc power.
    a. true   _____
    b. false

38. A large 25-hp, direct-current series motor will not operate satis-
    factorily on an ac power source.
    a. true   _____
    b. false

GLOSSARY

ACROSS-THE-LINE. Method of motor starting which connects the motor directly to the supply line on starting or running; also called *full voltage control.*

ALTERNATING CURRENT (ac). A current which alternates regularly in direction. Refers to a periodic current with successive half waves of the same shape and area.

ALTERNATOR. A machine used to generate alternating current by rotating conductors through a magnetic field; an alternating current generator.

ALTERNATOR PERIODIC TIME RELATIONSHIP. The phase voltages of two generators running at different speeds.

ALTERNATORS PARALLELED. Alternators are connected in parallel whenever the power demand of the load circuit is greater than the power output of a single alternator.

AMBIENT TEMPERATURE. The temperature surrounding a device.

AMORTISSEUR WINDING. Consists of copper bars embedded in the cores of the poles of a synchronous motor. The copper bars of this special type of squirrel-cage winding are welded to end rings on each side of the rotor; used for starting only.

ARMATURE. A cylindrical, laminated iron structure mounted on a drive shaft; contains the armature winding.

ARMATURE WINDING. Wiring embedded in slots on the surface of the armature; voltage is induced in this winding on a generator.

AUTOMATIC COMPENSATORS. Motor starters that have provisions for connecting three-phase motors automatically across 50%, 65%, 80%, and 100% of the rated line voltage for starting, in that order after preset timing.

AUTOTRANSFORMER. A transformer in which a part of the winding is common to both the primary and secondary circuits.

AUXILIARY CONTACTS. Contacts of a switching device in addition to the main circuit contacts; auxiliary contacts operate with the movement of the main contacts; electrical interlocks.

BRANCH CIRCUIT. The circuit conductors between the final overcurrent device protecting the circuit and the power outlet.

BRUSHLESS EXCITATION. The commutator of a conventional direct-connected exciter of a synchronous motor is replaced with a three-phase, bridge-type, solid-state rectifier.

BRUSHLESS EXCITER. Solid-state voltage control on an alternator, providing dc necessary for the generation of ac.

BUS. A conducting bar, of different current capacities, usually made of copper or aluminum.

BUSWAY. A system of enclosed power transmission that is current and voltage rated.

CAPACITOR. A device made with two conductive plates separated by an insulator or dielectric.

CENTRIFUGAL SWITCH. On single-phase motors, when the rotor is at normal speed, centrifugal force set up in the switch mechanism causes the collar to move and allows switch contacts to open, removing starting winding.

CIRCUIT BREAKER. A device designed to open and close a circuit by nonautomatic means and to open the circuit automatically on a predetermined overcurrent without injury to itself when properly applied within its rating.

COGENERATING PLANTS. Diesel powered electric generating sets which are designed to recapture and use the waste heat both from their exhaust and cooling systems.

COMMUTATOR. Consists of a series of copper segments which are insulated from one another and the mounting shaft; used on dc motors and generators.

CONDUCTOR. A device or material that permits current to flow through it easily.

CONDUIT PLAN. A diagram of all external wiring between isolated panels and electrical equipment.

CONTACTOR. An electromagnetic device that repeatedly establishes or interrupts an electric power circuit.

CONTROLLER. A device or group of devices that governs, in a predetermined manner, the delivery of electric power to apparatus connected to it.

CURRENT. The rate of flow of electrons; measured in amperes.

DC EXCITER BUS. A bus from which other alternators receive their excitation power.

DEFINITE TIME. A predetermined time lapse.

DELTA CONNECTION. A circuit formed by connecting three electrical devices in series to form a closed loop; used in three-phase connections.

DIODE. A two-element device that permits current to flow through it in only one direction.

DIRECT CURRENT (dc). Current that does not reverse its direction of flow. A continuous nonvarying current in one direction.

DISCONNECTING SWITCH. A switch that is intended to open a circuit only after the load has been thrown off by some other means. It is not intended to be opened under load.

DRUM SWITCH. A manually operated switch having electrical connecting parts in the form of fingers held by spring pressure against contact segments or surfaces on the periphery of a rotating cylinder or sector.

DUAL VOLTAGE MOTORS. Motors designed to operate on two different voltage ratings.

DUTY CYCLE. The period of time in which a motor can safely operate under a load. *Continuous* means that the motor can operate fully loaded 24 hours a day.

DYNAMIC BRAKING. Using a dc motor as a generator, taking it off the supply line and applying an energy dissipating resistor to the armature. Dynamic braking for an ac motor is accomplished by disconnecting the motor from the line and connecting dc power to the stator winding.

EFFICIENCY. The efficiency of all machinery is the ratio of the output to the input. Efficiency = output/input.

ELECTRIC CONTROLLER. A device, or group of devices, which governs, in some predetermined manner, the electric power delivered to the apparatus to which it is connected.

ELEMENTARY DIAGRAM (Ladder Diagram, Schematic Diagram, Line Diagram). Represents the electrical control circuit in the simplest manner. All control devices and connections are shown as symbols located between vertical lines that represent the source of control power.

EMERGENCY GENERATOR SYSTEM. A generating set which functions as a power source in a health care facility, such as a hospital; a standby power system. In addition to lighting, the loads supplied are essential to life and safety.

ENGINE-DRIVEN GENERATING SETS. Generators with prime movers of diesel or gasoline engines, or natural gas, and the like.

EXCITER. A dc generator that supplies the magnetic field for an alternator.

FEEDER. All circuit conductors between the service equipment, or the generator switchboard of an isolated plant, and the final branch-circuit overcurrent device.

FIELD DISCHARGE SWITCH. Used in the excitation circuit of an alternator. Controls (through a resistor) the high inductive voltage created in the field coils by the collapsing magnetic field.

FLUX. Magnetic field; magnetism.

FREQUENCY. Cycles per second or hertz.

FUSE. An overcurrent protective device with a circuit opening fusible part that is heated and severed by the passage of overcurrent through it.

GEAR MOTOR. A self-contained drive made up of a ball bearing motor and a speed reducing gear box.

GROUNDED. Connected to earth or to some conducting body that serves in place of earth.

GROWLER. An instrument consisting of an electromagnetic yoke and winding excited from an ac source; used to locate short-circuited motor coils.

HERTZ. The measurement of the number of cycles of an alternating current or voltage completed in one second.

JOGGING. The quickly repeated closure of a controller circuit to start a motor from rest for the purpose of accomplishing small movements of the driven machine.

LEGALLY REQUIRED STANDBY GENERATING SYSTEMS. Those systems required by municipal, state, federal or other codes or government agency having jurisdiction.

MAINTAINING CONTACT. A small contact in the control circuit used to keep a coil energized; usually actuated by the same coil; also known as a holding contact or an auxiliary contact.

MECHANICAL INTERLOCK. A mechanical interlocking device is assembled at the factory between forward and reverse motor starters and multispeed starters; it locks out one starter at the beginning of the stroke of either starter to prevent short circuits and burnouts by the accidental closure of both starters simultaneously.

MEGOHMMETER (MEGGER®). An electrical instrument used to measure insulation resistance.

MEGOHMS. A unit of resistance equal to 1,000,000 ohms.

MOTOR CIRCUIT SWITCH (Externally Operated Disconnect Switch, EXO). Motor branch circuit switch rated in horsepower. Usually contains motor starting protection; safety switch.

MOTOR CONTROLLER. A device used to control the operation of a motor.

MOTOR STARTER. A device used to start and/or regulate the current to a motor during the starting period. It may be used to make or break the circuit and/or limit the starting current. It is equipped with overload protection devices, such as a contactor with overload relays.

MULTIMETER. Electrical instrument designed to measure two or more electrical quantities.

NEC. National Electrical Code.

NONSALIENT ROTOR. A rotor that has a smooth cylindrical surface. The field poles (usually two or four) do not protrude above this smooth surface.

NORMAL FIELD EXCITATION. The value of dc field excitation required to achieve unity power factor in a synchronous motor.

NORMALLY OPEN and **NORMALLY CLOSED.** When applied to a magnetically operated switching device, such as a contactor or relay, or to the contacts of these devices, these terms signify the position taken when the operating magnet is de-energized, and with no external forces applied. The terms apply to the nonlatching types of devices only.

OHMMETER. An instrument used to measure resistance.

OIL (IMMERSED) SWITCH. Contacts of a switch that operate in an oil bath tank. Switch is used on high voltages to connect or disconnect a load. Also known as an oil circuit breaker.

OVERLOAD. Operation of equipment in excess of normal, full load rating, or of a conductor in excess of rated ampacity which, when it persists for a sufficient length of time, would cause damage or dangerous overheating.

OVERLOAD PROTECTION (Running Protection). Overload protection is the result of a device that operates on excessive current, but not necessarily on a short circuit, to cause the interruption of current flow to the device governed. Usually consists of thermal overload relay units inserted in series with the conductors supplying the motor.

PLUGGING. Braking a motor by reversing the line voltage or phase sequence; motor develops a retarding force; a quick stop.

PLUGGING RELAY. A device attached to a motor shaft to accomplish plugging; switches reversing starter to establish counter torque which brings the motor to a quick standstill before it begins to rotate in the reverse direction.

POLARITY. The characteristic of a device that exhibits opposite quantities, such as positive and negative, within itself.

POLE. The north or south magnetic end of a magnet; a terminal of a switch; one set of contacts for one circuit of main power.

POLYPHASE. An electrical system with the proper combination of two or more single-phase systems.

POWER FACTOR. The ratio of true power to apparent power. A power factor of 100% is the best electrical system.

PREVENTIVE MAINTENANCE. Periodic inspections to prevent serious damage to machinery by locating potential trouble areas; preventing breakdowns rather than repairing them.

PUSHBUTTON. A master switch; manually operated plunger or button for an electrical actuating device; assembled into push-button stations.

RACEWAY. A channel or conduit designed expressly for holding wires, cables, or busbars.

RATING. The rating of a switch or circuit breaker includes: the maximum current, voltage of the circuit on which it is intended to operate, the normal frequency of the current, and the interrupting tolerance of the device.

RECTIFIER. A device that converts alternating current (ac) into direct current (dc).

RELAY. Used in control circuits; operated by a change in one electrical circuit to control a device in the same circuit or another circuit.

REMOTE CONTROL. Controls the function initiation or change of an electrical device from some remote location.

RESISTANCE STARTER (Primary Resistance Starter). A controller to start a motor at a reduced voltage with resistors in the line on start.

R/min (or RPM). Speed in revolutions per minute.

ROTOR. The revolving part of an ac motor or alternator.

SALIENT FIELD ROTOR. Found on three-phase alternators and synchronous motors; field poles protrude from the rotor support structure. The structure is of steel construction and commonly consists of a hub, spokes, and a rim. This support structure is called a spider.

SELSYN. Abbreviation of the words self-synchronous. Selsyn units are special ac motors used primarily in applications requiring remote control. These units are also referred to as synchros.

SEMICONDUCTORS. Materials which are neither good conductors nor good insulators. Certain combinations of these materials allow current to flow in one direction but not in the opposite direction.

SEPARATE CONTROL. The coil voltages of a relay, contactor or motor starter are separate or different from those at the switch contacts.

SEPARATELY-EXCITED FIELD. The electrical power required by the field circuit of a dc generator may be supplied from a separate or outside dc supply.

SERVICE FACTOR. An allowable motor overload; the amount of allowable overload is indicated by a multiplier which, when applied to a normal horsepower rating, indicates the permissible loading.

SILICON-CONTROLLED RECTIFIER (SCR). A four-layer semiconductor device that is a rectifier. It must be triggered by a pulse applied to the gate before it will conduct electricity.

SINGLE PHASE. A term characteristic of a circuit energized by a single alternating emf. Such a circuit is usually supplied through two wires.

SLIP. In an induction motor, slip is the difference between the synchronous speed and the rotor speed, usually expressed as a percentage.

SLIP RINGS. Copper or brass rings mounted on, and insulated from, the shaft of an alternator or wound rotor induction motor; used to complete connections between a stationary circuit and a revolving circuit.

SOLENOID. An electromagnet used to cause mechanical movement of an armature, such as a solenoid valve.

SOLID STATE. As used in electrical-electronic circuits, refers to the use of solid materials as opposed to gases, as in an electron tube. It usually refers to equipment using semiconductors.

SPEED CONTROL. Refers to changes in motor speed produced intentionally by the use of auxiliary control, such as a field rheostat or automatic equipment.

SPEED REGULATION. The variation in the speed of a motor caused by a change in the load. A motor whose speed remains practically constant from no load to full load is considered to have good speed regulation.

SPLIT PHASE. A single-phase induction motor with auxiliary winding, displaced in magnetic position from, and connected parallel to, the main winding.

STANDBY POWER GENERATING SYSTEM. Alternate power system for applications such as heating, refrigeration, data processing, or communications systems where interruption of normal power would cause human discomfort or damage to a product in manufacture.

STARTING CURRENT. The surge of amperes of a motor upon starting.

STARTING PROTECTION. Overcurrent protection is provided to protect the motor installation from potential damage due to short circuits, defective wiring, or faults in the motor controller or the motor windings. The starting protection may consist of a motor disconnect switch containing fuses.

STATOR. The stationary part of a motor or alternator; the part of the machine that is secured to the frame.

SYNCHRONOUS ALTERNATORS. Frequencies, voltages, and instantaneous ac polarities must be equal when paralleling machines.

SYNCHRONOUS CAPACITOR. A synchronous motor operating only to correct the power factor and not driving any mechanical load.

SYNCHRONOUS MOTOR. A three-phase motor (ac) which operates at a constant speed from a no load condition to full load; has a revolving field which is separately excited from a direct current source; similar in construction to a three-phase ac alternator.

SYNCHRONOUS SPEED. The speed at which the electromagnetic field revolves around the stator of an induction motor. The synchronous speed is determined by the frequency (hertz) of the supply voltage and the number of poles on the motor stator.

SYNCHROSCOPE. An electrical instrument for synchronizing two alternators.

TACHOMETER. An instrument used to check the speed of a motor or machine.

THREE PHASE. A term applied to three alternating currents or voltages of the same frequency, type of wave, and amplitude. The currents and/or voltages are one-third of a cycle (120 electrical time degrees) apart.

THREE-PHASE SYSTEM. Electrical energy originates from an alternator which has three main windings placed 120 degrees apart. Three wires are used to transmit the energy.

TORQUE. The rotating force of a motor shaft produced by the interaction of the magnetic fields of the armature and the field poles.

TRANSFER SWITCHES. Switches to transfer, or reconnect, the load from a preferred or normal electric power supply to the emergency power supply.

TRANSFORMER. An electromagnetic device that converts voltages for use in power transmission and operation of control devices.

TRANSFORMER BANK. When two or three transformers are used to step down or step up voltage on a three-phase system.

WHEATSTONE BRIDGE. Circuit configuration used to measure electrical qualities such as resistance.

WIRING DIAGRAM. Locates the wiring on a control panel in relationship to the actual location of the equipment and terminals; made up of a method of lines and symbols on paper.

WOUND ROTOR INDUCTION MOTOR. An ac motor consisting of a stator core with a three-phase winding, a wound rotor with slip rings, brushes and brush holders, and two end shields to house the bearings that support the rotor shaft.

WYE CONNECTION (Star). A connection of three components made in such a manner that one end of each component is connected. This connection generally connects devices to a three-phase power system.

INDEX